NUTRITION, IMMUNITY, AND INFECTION
Mechanisms of Interactions

Annapurna, the goddess of food and plenty specially revered in the villages of India. From a mud mural. The importance of dietary intake and physical and mental health is discussed in ancient Indian scriptures written several centuries B.C. (Courtesy of India Tourism Development Corporation.)

NUTRITION, IMMUNITY, AND INFECTION
Mechanisms of Interactions

R. K. CHANDRA
Memorial University of Newfoundland
St. John's, Newfoundland, Canada

P. M. NEWBERNE
Massachusetts Institute of Technology
Cambridge, Massachusetts

PLENUM PRESS • NEW YORK AND LONDON

Library of Congress Cataloging in Publication Data

Chandra, R   K
  Nutrition, immunity, and infection.

  Bibliography: p.
  Includes index.
  1. Immune response. 2. Nutrition. 3. Infection. 4. Immunological deficiency syndromes—Nutritional aspects. I. Newberne, Paul Medford, 1920-   joint author. II. Title. [DNLM: 1. Nutrition disorders—Complications. 2. Immunity. 3. Diseases—Etiology. [QZ185 G456n]
  QR186.C48   616.9   77-21209
  ISBN 0-306-31058-9

First Printing – December 1977
Second Printing – June 1979

© 1977 Plenum Press, New York
A Division of Plenum Publishing Corporation
227 West 17th Street, New York, N.Y. 10011

All rights reserved

No part of this book may be reproduced, stored in a retrieval system, or transmitted, in any form or by any means, electronic, mechanical, photocopying, microfilming, recording, or otherwise, without written permission from the Publisher

Printed in the United States of America

To our families

FOREWORD

I welcome the privilege of writing some words of introduction to this important book. Its authors have been courageous in bringing together in one text a triad of topics that cover such large tracts of biomedical sciences as epidemiology, biochemistry, immunology, and clinical medicine. Malnutrition and infection are known to be closely linked, the one promoting the other. The adaptive immune system forms a part of the link since it is responsible for a good deal of defense against infection, and it may be affected adversely by malnutrition and indeed by infection itself. Knowledge in this complex field is of great potential importance because malnutrition and infection are such dominant features of the ill-health of many of the world's underprivileged people.

As this book shows, there is no lack of technical facets for study. There are now so many components of the immune response which can be measured or assessed and so many aspects of nutritional biochemistry which can be studied that the problem is to select what to study and where to begin. Moreover, the great number of variables in the nature of nutritional deficiencies, in types of infections or multiple infections and in the genetic, environmental, and social background of the affected people, all combine to make interpretation and application of findings a speculative business. Descriptions of cause and effect must usually be provisional rather than definitive. There are, quite rightly, so many occasions in this book when a conditional verb must be used that research workers could be tempted to ask whether these topics can indeed be integrated in a systematic and useful way.

That is no reason for not trying. The biomedical scientist trying to apply his knowledge and skills to the improvement of health in the community cannot expect to find controlled conditions comparable to those that can be achieved in the test tube or the inbred mouse. Some consistency has already emerged in this complex field. In malnourished children, the T cell system—thymus, T lymphocyte, and cell-mediated response—are regularly impaired, although amounts of immunoglobulin are not reduced and may even be raised. Antibody formation is normal or reduced, perhaps (note the conditional clause) due to defective T cell function. Such findings imply, for example, a relationship between nutritional status and the effectiveness of vaccines. No doubt all nutritional problems could be solved by food, but unless we are so sanguine as to imagine that an optimal diet will in the near future be available for all, it would be well to consider possible strategies to minimize the adverse effects of malnutrition.

Immunization with present vaccines, and the new ones to come, might be more effective if carried out under nutritionally favorable circumstances as, for example, during a season when food was more plentiful or after a period of dietary supplementation. Alternatively, it may be possible to design the vaccine, and, in particular, its adjuvant, to stimulate an effective T cell response even in malnourished subjects.

Much of the research described in this book concerns the essential laboratory base upon which further advances can be made. But even the most elegant triumphs in this field will ring hollow if their relevance to the control of human disease cannot be assessed or realized. There is now, more than ever, a need to link the more fundamental laboratory and clinical studies with observations on man. Longitudinal epidemiological studies on infectious diseases that seek to assess the significance of nutrition and immunological changes alongside other factors such as genetics and social and economic circumstances are one powerful approach that has already been exploited in some areas. This book will be a valuable source of knowledge and ideas for such studies.

DAVID S. ROWE

Special Programme for Research and Training in Tropical Diseases
World Health Organization
Geneva, Switzerland

PREFACE

The intimate and complex relationship between diet and health finds mention in the ancient scriptures of India, in the treatises of Chinese and Roman medicine, and in the Corpus Hippocratum. It is sad that in spite of phenomenal advances in agricultural techniques and medical sciences, we still see in 1977 the struggle between diminishing food supply and increasing population growth in its starkest form. Undernutrition, often combined with the intertwined problem of infection, continues to threaten the health and survival of the majority of the world's population, particularly infants and children. If nutritional deficiency and susceptibility to infectious illness are a conjugate pair, it is important to understand the pathogenetic processes involved.

Recent advances in immunological concepts and techniques have stimulated many studies in the general field of nutrition–immunocompetence–infection interactions. Several noncellular factors and the number, morphology, and function of many cell types have been found to be altered in nutritional deficiency states. The studies have provided an insight into the intricacies of Nature's methods of preferential synthesis of cells, proteins, and enzymes required for host defense, at the expense of expendable elements. Nutritional modulation of immunity may be an important determinant of morbidity and mortality associated with a variety of disease processes. Interestingly, undernutrition as well as overnutrition can alter immune responsiveness. Thus "optimum nutrition" is the key phrase for dietary influences to keep immune response within normal limits. Questions arising from observations in man have led to controlled experimentation in laboratory animals. It is unlikely that the results of animal experiments can be directly extrapolated to the human situation. However, such fundamental observations point the way to the interesting possibility of dietary manipulation of immune response, which may determine susceptibility or resistance to many diseases.

The data generated have to be interpreted with caution. Many aspects of the immune response have been studied, often in isolation from other parameters and seemingly out of context with other critical environmental variables which influence the occurrence of infectious illness. In the feverish excitement of research, it is easily forgotten that in complex biological systems the total effect of a series of components is not necessarily the sum of their individual effects. Differences in the nutritional status of the subjects evaluated, variations in the techniques used, the presence or absence of complicating infections and stress hormonal and metabolic processes, make it difficult to compare and collate observations.

In this interpretative monograph, we have attempted to summarize and analyze critically the reported information, to bring out the controversies, and to point out the lacunae where further data are required. We have drawn upon the work of others and of our own, and have included some previously unpublished observations. We have asked many questions which cannot at present be answered. The inevitability of increase in scientific knowledge cannot be denied. It is our fervent hope that the efforts of many research workers and health professionals will lead to solutions with the ultimate objective of preventing or alleviating suffering generated by malnutrition and infection.

We wish to acknowledge the excellent assistance of Rose Marie Puddicombe, Louise Kittridge, Rosemary Burklin, Clifford George, Gordon King, and Jim Thistle, in the preparation of the manuscript.

R. K. CHANDRA
P. M. NEWBERNE

CONTENTS

1. INTRODUCTION 1

2. MECHANISMS OF HOST DEFENSE 11

    2.1. Ontogenetic Development of Antigen-Specific
         Immunity 12
         2.1.1. Thymus-Dependent Cell-Mediated
                Immunity 14
         2.1.2. Immunoglobulins and Antibodies . . 16
    2.2. Phagocytes 21
         2.2.1. Monocytic Phagocyte 24
    2.3. Complement System 25
    2.4. Other Factors of Host Resistance 28
    2.5. Concluding Remarks 28

3. ASSESSMENT OF NUTRITIONAL STATUS . . 31

    3.1. Energy–Protein Undernutrition 31
    3.2. Vitamin and Mineral Deficiencies 38
    3.3. Fetal Malnutrition 38
    3.4. Concluding Remarks 40

4. INFECTIONS IN UNDERNOURISHED
   INDIVIDUALS 41

5. **NUTRITIONAL, METABOLIC, AND IMMUNOLOGIC EFFECTS OF INFECTION** . . 47

    5.1. Nutritional Consequences of Infection . . . 47
        5.1.1. Proteins 50
        5.1.2. Carbohydrates 51
        5.1.3. Lipids 52
        5.1.4. Minerals and Electrolytes 52
        5.1.5. Vitamins 53
    5.2. Metabolic and Hormonal Responses during Infection 53
    5.3. Immune Responses in Infection 56
        5.3.1. Cell-Mediated Immunity 57
        5.3.2. Immunoglobulins and Antibodies . . 61
        5.3.3. Phagocytes 63

6. **IMMUNOCOMPETENCE IN UNDERNUTRITION** 67

    6.1. Lymphoid Tissues 67
        6.1.1. Thymus 68
        6.1.2. Lymph Nodes 71
        6.1.3. Spleen 71
        6.1.4. Gut-Associated Lymphoid Aggregates 71
        6.1.5. Pathogenesis 74
    6.2. Cell-Mediated Immunity 74
        6.2.1. Leukocyte Counts 74
        6.2.2. Lymphocyte Subpopulations . . . 75
        6.2.3. Delayed Hypersensitivity 80
        6.2.4. Lymphocyte Proliferation 85
        6.2.5. Methodological and Pathogenetic Considerations 87
    6.3. Immunoglobulins and Antibodies 89
        6.3.1. γ-Globulin Concentration and Metabolism 89
        6.3.2. Immunoglobulin Levels and Turnover 90
        6.3.3. Serum Antibody Response 96

		6.3.4.	Secretory Antibody Response	101
6.4.			Complement System	104
6.5.			Phagocytes	110
		6.5.1.	Number and Morphology	110
		6.5.2.	Mobilization	112
		6.5.3.	Chemotaxis and Inflammatory Response	112
		6.5.4.	Metabolism	113
		6.5.5.	Opsonization, Phagocytosis, and Bactericidal Capacity	116
6.6.			Lysozyme	121
6.7.			Iron-Binding Proteins	123
6.8.			Tissue Integrity	124
6.9.			Interferon	124
6.10.			Other Factors	125
		6.10.1.	Gut Microflora	125
		6.10.2.	Hormones	126
		6.10.3.	Miscellaneous	126

7. **INTERACTIONS OF NUTRITION, INFECTION, AND IMMUNE RESPONSE IN ANIMALS** . . . 127

7.1.			General Considerations	127
		7.1.1.	Introduction	127
		7.1.2.	Species Variations in Dietary Needs	128
		7.1.3.	Synergism and Antagonism	131
7.2.			Nutrition–Infection Interactions	132
		7.2.1.	Protein	132
		7.2.2.	Fat	150
		7.2.3.	Carbohydrates	152
		7.2.4.	Vitamins	152
		7.2.5.	Calories	171
		7.2.6.	Minerals	172
		7.2.7.	Effects of Excessive Nutrition	177
		7.2.8.	Vitamin A	180
7.3.			Summary	180

8. BIOLOGICAL IMPLICATIONS 181

    8.1. Infection-Related Morbidity and Mortality . . 181
    8.2. Postoperative Sepsis 182
    8.3. Parenteral Hyperalimentation and Infection . 182
    8.4. Intergenerational Effects of Undernutrition, Impaired Immunocompetence, and Infection . 183
    8.5. Nutritional Deficiency, Immunopathologic Disease, and Aging 185
    8.6. Cancer 187
    8.7. Autoimmunity and Allergy 190
    8.8. Food Antibodies 190
    8.9. Prophylactic Immunization 192
    8.10. Immunopotentiation in Management of Malnutrition–Infection Syndrome 194
    8.11. Obesity 196

9. FUTURE RESEARCH NEEDS 199

REFERENCES 203

INDEX 235

NUTRITION, IMMUNITY, AND INFECTION
Mechanisms of Interactions

1

INTRODUCTION

The important role of nutritional deficiency as a contributor to childhood mortality particularly from infectious disease, as a conditioning factor in the complex mosaic of many diseases, and as a hurdle to socioeconomic advancement, is widely recognized. Improvements in nutrition, together with better hygiene and immunization, can take most of the credit for decreasing death rate from infectious disease and for the longer expectation of life in industrialized countries. This change in the mortality pattern came before the development of antibiotics and modern medical techniques. In developing countries, gross life-threatening malnutrition in the shape of marasmus and kwashiorkor continues to be rampant. These dramatic syndromes represent the clinical end points of nutritional pathology, and form the tip of the massive iceberg of undernutrition. More than 100 million preschool infants and children suffer from moderate-severe malnutrition in the world. And for one case of kwashiorkor or marasmus, there are at least 100 with mild to moderate deficiencies of one or more nutrients. Recent studies in the Americas have revealed a surprisingly high incidence of nutritional problems. The Pan American Health Organization survey of mortality patterns carried out in Central and South America showed malnutrition and infections to be the most serious health problems in children, as primary or more often as secondary factors in deaths (Puffer and Serrano, 1973). Fifty-seven percent of the children who died under 5 years of age revealed signs of undernutrition, intrauterine and/or after birth, as either the underlying or an associated cause of mortality. The death rate from infectious diseases, mainly diarrhea and measles, was found to be 23%. Random population surveys in the United States (Department of Health, Education and Welfare, 1972) and in Canada (Nutrition Canada, 1973) have revealed a high prevalence of nutritional problems particularly among the impoverished segments of society. An examination of 300 randomly picked preschool children of poor black families in Memphis showed that about one-sixth were below the third percentile of standards of weight and height, one-fourth were anemic, 27% had retarded skeletal development, and 44% had low levels of vitamin A (Zee *et al.,* 1970).

Clinicians have long observed that undernutrition predisposes the host to the risk of acquired infection and that the course, frequency of complications, severity, and mortality of the infectious illness are augmented. It is likely that this is the result, in part, of impaired immunocompetence secondary to nutritional deficiency. Infection in turn frequently worsens the nutritional status, often precipitating overt symptoms and signs, and causes immunosuppression. A variety of complex pathogenetic mechanisms probably underlie these multicornered interactions (Fig. 1.1).

There is an intimate relationship between nutritional status, immune response, and infection (Fig. 1.2). The common concurrent existence of malnutrition and infection may symbolize a pathophysiologic interaction between the two, to produce effects including immunosuppression that are more than the summed result expected from the two diseases acting singly, the phenomenon of synergism (Scrimshaw *et al.,* 1968). Or an antagonism may exist. Alternatively, the frequent coexistence of malnutrition and infection may simply indicate the occurrence of common causative factor(s) in the same ecosystem. In most industrialized countries, economic affluence, availability of food in sufficient amounts, immunizations, and improvement in sanitation have led to the decline of malnutrition as well as infectious illnesses. In the economically less privileged nations, however, the opportunities for combined mutually aggravating effects of undernutrition and infection continue to prevail and to pose a threat to the health of the majority of their populations, most particularly young children under 5 years of age.

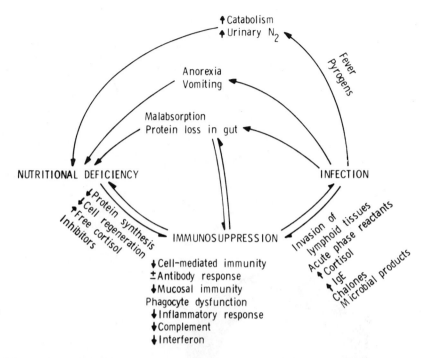

Figure 1.1. Mechanisms of interactions between infection, nutritional deficiency, and immunosuppression.

In the continuous struggle between the host and the pathogen, resistance ability of the former and virulence of the latter are the key determinants (Fig. 1.3). If the organism is highly pathogenic, for example measles virus, disease is the invariable result in the nonimmunized person. In the case of the relatively avirulent organism, for example *Candida albicans*, generally no disease ensues, and "immunity" develops. In the vast majority of instances, however, both these forces are of variable intermediate strength, and in such a situation factors modifying one or the other may tip the balance to morbidity and mortality, or to symptomatic infection with complete recovery. Nutrition is one of the critical determinants in this balance (Chandra, 1976b).

When nutritional deficiency and infection coexist, the former is often chronic and precedes the latter acute process. Infectious illness is likely to cause the greatest havoc during periods of rapid

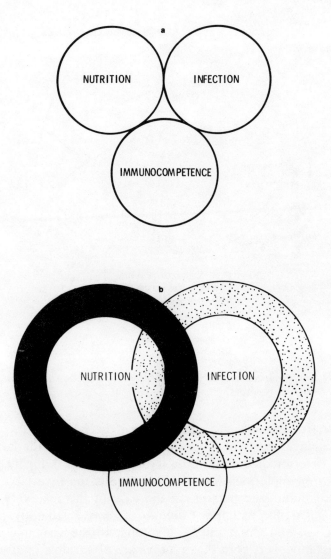

Figure 1.2. Interactions between nutritional status, immunocompetence, and infection. Abnormality in one field affects the other two. (a) Nutritional deficiency impairs immunocompetence and increases the frequency and severity of infection. Infectious illness is associated with negative nitrogen balance, may precipitate overt malnutrition, and depresses immune responses. Primary immunodeficiency states are characterized by failure to thrive and a variable susceptibility to infection. (b) Undernutrition and infection often coexist and augment each other and impair immunocompetence to a variable extent.

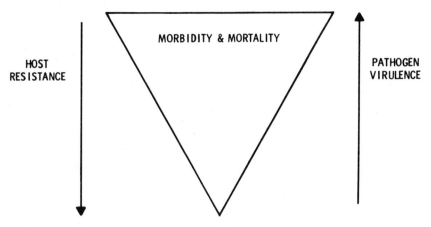

Figure 1.3. The magnitude of morbidity and mortality in infectious disease is the net balance of the opposing forces of pathogen virulence and host resistance. Organisms such as measles virus produce symptomatic disease in virtually all nonimmune individuals. Other agents such as *Candida albicans* exist in symbiosis with man until impairment of host immunity allows it to produce opportunistic infection.

growth with high physiological demands of nutrients: intrauterine life, the first two years of life, adolescence, pregnancy and lactation, and old age. The frequent occurrence of elevated cord blood IgM in poor communities reflects the high incidence of prenatal infections. The duration and severity of the preceding undernutrition and the length and severity of superadded infection determine the biological importance of the resulting interaction. In an infant getting a marginally adequate diet, heavy parasitic burden and closely spaced infections invariably retard growth or produce an actual loss of weight (Fig. 1.4). Case incidence of common infections, such as gastroenteritis, respiratory illnesses, and tuberculosis, is increased and infections contribute to augmented preschool morbidity and mortality.

The mutually augmenting effects of malnutrition and infection are seen not only in individuals with gross energy-protein undernutrition, but also in those with deficiency of individual nutrients, such as iron, folates, pyridoxine, etc. The clinical association of anemia and infections is known, but one may be a cause or

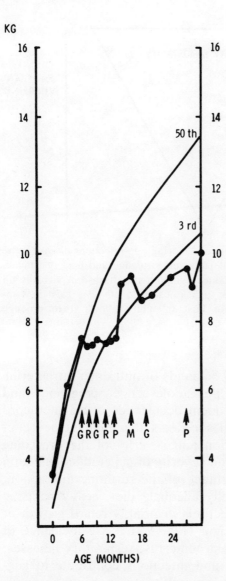

Figure 1.4. Impact of infectious illness on growth. The infant's weight in the first six months of life was on the 50th percentile of standard. Recurrent episodes of gastroenteritis G, respiratory infection R, pneumonia P, and measles M slowed the rate of weight gain and occasionally produced an actual weight loss, and brought down the child's growth curve to below the 3rd percentile.

consequence of the other. There is a higher than expected frequency of clinical or subclinical iron deficiency in patients with chronic mucocutaneous candidiasis (Higgs and Wells, 1973) and recurrent herpes virus infection (Chandra *et al.*, 1977d). Clinical surveys estimating the incidence of infection in relationship to the individual's iron status have yielded variable results (Chandra, 1976a). Much of the data is difficult to interpret because of inadequate sampling, controls, follow-up attrition, and statistical methodology. In a survey of construction and plantation workers in Indonesia, Basta and Churchill (1974) reported a higher prevalence of acute and chronic infections in the iron-deficient anemic group compared with nonanemic controls. Some studies report a reduction in the incidence of respiratory infections and gastroenteritis in infants given iron supplements, whereas other studies have failed to confirm such an effect (Pearson and Robinson, 1976; Chandra, 1977b).

Let us turn the picture around and relook at the very basis of the unchallenged epidemiologic opinion that infections are commoner and more severe in the malnourished. The clinical impression of health workers dealing with underfed populations, the comparison of morbidity and mortality data obtained from undernourished and well-nourished groups often far removed from each other geographically and ethnically, and change in such data after nutritional supplementation, has supported the concept of synergism between malnutrition and infection. However, in many such studies, the other critical variables of infection frequency, *viz.*, sanitation, personal hygiene, overcrowding, family size, were not taken into consideration.

On the other hand, there are many bits of circumstantial and direct evidence which, if pieced together, dispute the concept that lack of dietary nutrients makes an individual more susceptible to infection, and that in fact starvation exerts an overall beneficial effect for man in the continuous ecological struggle between human beings and their microbial predators in the environment.

Experimental studies suggest that host resistance to many viral infections and to tumors is enhanced by undernutrition. The pathologic conditions examined in this fashion include vaccinia, poliomyelitis, foot-and-mouth disease, lymphocytic chorio-meningitis, and sarcoma. In cattle, there is direct relationship between food supply, nutritional status, and incidence of foot-and-mouth disease. Lean animals with limited food intake have the lowest incidence of the disease. Similar data is available for hibernating or starving hedgehogs. Cooper *et al.* (1974) reviewed their data showing that chronic protein deficiency in mice did not alter either the primary or the secondary antibody response to immunization with *Brucella abortus,* destruction of *Listeria monocytogenes* was facilitated, and allograft rejection was hastened. Contrarily, streptococcal infection was enhanced. In protein-deficient animals, Jose and Good (1971, 1973) showed enhanced cell-mediated immune destruction of allogeneic, syngeneic, and autochthonous tumors. Negative data must, however, be interpreted with extreme caution, since several critical factors including the dependence of the species for the nutrient being investigated may exert important modulating influence. The majority of studies on nutritionally deprived laboratory animals observed that morbidity and mortality following infectious challenge are increased. Several animal species, many nutrients, and a variety of pathogenic organisms including bacteria, viruses, fungi, rickettsia, and parasites, have been studied. Scrimshaw *et al.* (1968) reviewed the extensive experimental data on the interactions between malnutrition and infection and concluded that the two factors generally aggravate each other, and only rarely does antagonism occur.

Murray and Murray (1977) have summarized the historic and epidemiologic information supporting the thesis that starvation suppresses infection and refeeding activates it. Susceptibility to epidemic diseases and mortality thereof were higher among the rich and well-fed, during the 1914–18 pandemic of influenza and among prisoners in the concentration camps of World War II. Well-fed prisoners in English jails suffered a higher morbidity and mortality, perhaps from infection. In anorexia nervosa, there is a surprisingly low incidence of infections. The seasonal periodicity of malaria in some tropical areas has been ascribed, in part, to the increased availability of food and presumably improved nutritional status after the monsoons. Outbreaks of malaria, brucellosis, and tuberculosis were observed when famine victims showed significant weight gain.

The complexities of environmental influences and host factors in determining susceptibility to and severity of infection defy simplistic analysis. On the one hand, there is the strong clinical impression of a synergistic interaction between nutritional deficiency and infection, possibly mediated through atrophy of the thymus and other lymphoid tissues, and impairment of specific and nonspecific immunologic and nonimmunologic protective responses. Nutritional therapy corrects these abnormalities, but perhaps not soon enough since many patients still die of complicating infections and other complications. On the other hand, there is some evidence in support of the thesis that calculated mild undernutrition and leanness may be an animal's greatest physical asset, producing longer life, fewer malignancies, reduced mortality from inherited susceptibility to autoimmune disease, and perhaps fewer infections. These biological paradoxes remain a riddle awaiting solution through more critical data.

2

MECHANISMS OF HOST DEFENSE

A complex set of structures and processes is involved in protecting individuals from infections, other foreign matter, and mutant cells developing *de novo* in the body, and their complications. These defense mechanisms may be specific for the invading microbe, or antigen-nonspecific; some are immunological, others nonimmunological (Table 2.1). Although individual populations of cells, proteins, and factors have been generally described and studied in isolation, there is an intimate interaction *in vivo* between different mechanisms of host defense. The immunologic orchestra plays in concert, rather than solo.

The functional significance of each component has been deduced from qualitative and quantitative primary immunity deficiency states, "the crucial experiments of Nature" (Good, 1973), and by correction by passive replacement or reconstitution therapy. The relative importance of different protective processes would depend upon the biologic characteristics of the infecting agent and the route of invasion. These defense mechanisms attempt to prevent colonization, systemic entry, and multiplication and dissemination of pathogens, and influence the consequent complications and final outcome: no symptoms, disease, recovery, or fatality.

The body's defense mechanisms may not always be beneficial. In many instances, the very processes that are mounted against invading pathogens may themselves cause tissue damage, clinical manifestations and complications associated with the disease. At the end of the 19th century, the Austrian pediatrician Clemens von Pirquet had postulated that an interaction of antibodies with the viruses of smallpox and measles might cause the skin eruptions characteristic of these diseases. These speculations that immune response might cause disease are fully borne out by numerous studies. The immunologic processes underlying local or systemic damage in disease have been categorized by Coombs and Gell (1975) into four basic types of reactions: Type I, anaphylactic or reagin-dependent; Type II, cytotoxic or cell stimulating; Type III, antigen–antibody or immune complexes; Type IV, delayed cell-mediated with release of lymphokines and/or development of cytotoxicity. A number of different processes may be going on at once in the same patient and even in the same lesion. If these immune responses are exaggerated or subdued, quantitatively or qualitatively, there is a greater likelihood of complications and an adverse ultimate outcome.

2.1. ONTOGENETIC DEVELOPMENT OF ANTIGEN-SPECIFIC IMMUNITY

The two-component concept of the cellular systems subserving specific immunological functions is now established. The pluripotent stem cell can be identified first of all in the yolk sac. Subsequently, the precursors of the major hematopoietic cells, including lymphocytes, granulocytes, and macrophages, can be found in the fetal liver and bone marrow. The ontogenetic development of the primodia of various lymphoid organs and their population by functional lymphoid cells has a characteristic pattern (Chandra, 1978b). Lymphocyte precursors ultimately develop into one of the two major subsets distinguished on the basis of function and cell surface markers: thymus-dependent T lymphocytes, and bone-marrow- or Bursa-dependent B lymphocytes (Lawton and Cooper, 1973). The development of these two defined but interdependent populations of immunocompetent cells is shown in Figure 2.1. The mature T and B lymphocytes are responsible for the two main components of antigen-specific protection, cell-mediated immunity, and the immunoglobulin–

Table 2.1
Host Defenses

I. Specific immunological responses
    Immunoglobulins and antibodies
        Serum
        Secretory
    Cell-mediated immunity
        T lymphocytes (helper, suppressor, cytotoxic, etc.)
        Killer cells

II. Nonspecific factors of resistance
    Skin and mucous membranes
    Mucus
    Visceral and ciliary movements
    Phagocytes
    Complement system
    Opsonic function
    Iron-binding proteins: transferrin, lactoferrin
    Interferon
    Lysozyme
    Febrile and metabolic responses

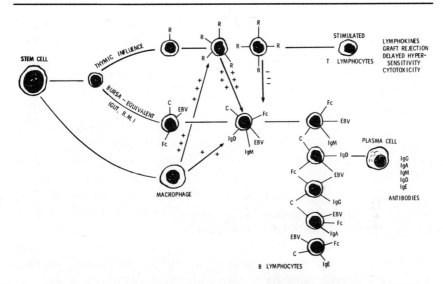

Figure 2.1. Development of the cells involved in antigenic-specific cell-mediated and humoral immunity. There is a complex interaction—+, helper or facilitatory; −, inhibitory—between T lymphocytes, B lymphocytes, and macrophages. Surface markers are progressively acquired. R = rosetting with sheep erythrocytes, C = complement, Fc = crystalline fragment of immunoglobulin, EBV = Epstein-Barr virus, B.M. = bone marrow.

antibody system, each of which may act in systemic sites or at mucosal surfaces.

2.1.1. Thymus-Dependent Cell-Mediated Immunity

Some of the lymphoid cells come under the functional influence of the thymus. The pattern and timetable of development of the thymus varies in different species. In man, the epithelial portion of the thymus has a common embryonic origin around the 6th week of gestation, from the 3rd and 4th branchial arches with the parathyroids and the thyroid. This embryologic feature explains the occurrence of hypocalcemic tetany, peculiar facies, and thymic aplasia in DiGeorge's syndrome. Around the 8th week of gestation, lymphocytes enter the thymus, where the cortical area is an active site of cell proliferation, acquisition of cell surface antigens, and functional maturation. It is possible that the generation of a diversity of specific antigen receptors takes place during thymic lymphopoiesis. The majority of the new cells die within the thymus (Mutsuyama *et al.*, 1966), a process which may be aimed at specialization, individual antigen recognition, and elimination of the "forbidden clones" directed against one's own tissues. The maturation effect of the thymic microenvironment is the result of soluble factors (? "thymosin," ? other thymic hormones) since it can be achieved by dialysate permeating across millipore filters. These humoral substances perhaps assist in the maturation of T cells in the peripheral tissues as well, and in the full expression of their function. The mature T lymphocytes move to the medullary core of the thymus and then leave the organ, to populate secondary lymphoid tissues such as the paracortical areas of lymph nodes, periarteriolar areas of the spleen, gut-associated lymphoid tissue, etc. Some of these T cells are long-lived and keep recirculating via the lymphatics, thoracic duct, and afferent venous drainage, perhaps for many years.

Mature human T lymphocytes are recognized by their ability to form nonimmune rosettes with sheep erythrocytes in the cold (Fig. 2.2), and their reaction with antisera raised against human thymocyte antigens and antisera against human brain cells. The electrokinetic behavior in free-flow electrophoresis, cell-surface molecular components (Mehrishi and Zeiller, 1974), membrane-

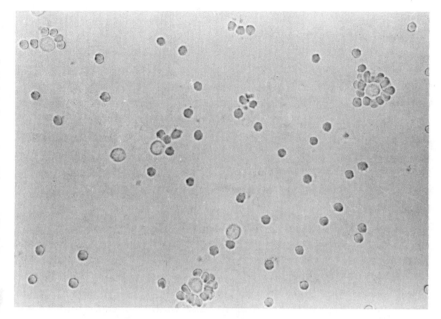

Figure 2.2. Rosette-forming human T lymphocytes.

bound acetylcholinesterase (Kutty et al., 1976) and α-naphthyl acetate esterase activity (Ranki et al., 1976) can also differentiate T lymphocytes from B cells. Small amount of immunoglobulin have been shown on the surface of T cells by extremely sensitive autoradiographic techniques and by electron microscopy. The exact function of this immunoglobulin and the nature of the antigen-recognition receptor are the subject of much current debate and work. Functionally, T lymphocytes may have a helper or suppressor effect on other T and B lymphocytes. They function through release of soluble factors (lymphokines) and direct cell–cell cytotoxicity, and participate in delayed hypersensitivity and homograft rejection.

In mice, a number of antigens have been demonstrated on the surface of T cells. Different combinations of antigens are observed in progressive stages of ontogenetic development, some being acquired initially only to be lost at a later stage, for example some Ly antigens, whereas others, for example theta (θ), being a permanent cell characteristic. Lymphocytes phenotyped on the basis of surface antigens have different functions (Medawar and Simpson, 1975; Janeway *et al.*, 1975).

There is convincing evidence to suggest that cellular immunity is functional early in gestation. It may be required to deal with maternal leukocytes which may provoke graft-versus-host reaction. Mixed leukocyte reaction can be demonstrated as early as the 10th week. A functionally adequate cellular immunity may help in limiting the extent of fetal infections and of contagiousness. On the other hand, the immunologic and inflammatory response to the invading pathogen may be responsible for the clinical manifestations, as exemplified by intrauterine syphilis infection.

2.1.2. Immunoglobulins and Antibodies

The immunoglobulins are produced by the B lymphocyte–plasma cell system. The early stages of development of cells involved in the synthesis and release of immunoglobulins have been studied extensively in the chicken, in which a well-demarcated cloacal Bursa of Fabricius serves as the central thymic equivalent for the B cell system. The high rate of cell division in this organ suggests that new specificities and antibody-combining sites are being continuously generated. A similar microenvironment suitable for induction probably exists in man but its anatomic localization has not been established. The gut-associated lymphoid tissue, particularly the appendix and Peyer's patches, and the bone marrow, is considered to be the mammalian homologue of the Bursa. In newborn rabbits (Cooper *et al.*, 1968) and in the human neonate (Chandra, 1977d), extensive bowel resection is associated with lower than normal levels of immunoglobulins. Recent evidence suggests that the bone marrow, liver, and spleen may be the Bursa-equivalent organs in mammals.

Cooper *et al.* (1972) have postulated a two-stage model for sequential ontogenetic development of B cells and differentiation of plasma cells. The first stage of clonal development involves the origin of antigen-reactive cells from stem cells. It is suggested that cells synthesizing IgM and expressing genes for the variable and constant (C_μ) regions give rise to cells producing IgG (C_γ), which in turn may switch to IgA (C_α) synthesis. This intraclonal differen-

has been supported by observations on
munity deficiency and cell-surface char-
ignant cells, and in chickens that were
antisera specific for one type of heavy
consists of clonal proliferation which
ies. It is set into motion by antigenic
s multiply and differentiate into immu-
a cells, which are end cells incapable of

recognized by the presence of surface
ch can be easily detected by staining
ted antihuman-immunoglobulins anti-
oscopy (Fig. 2.3). Autoradiography and
opy are alternate research tools for
most frequently employed technique is
ep red cells induced in the presence of
t. Other markers on the surface of
ptors for aggregated IgG, Fc, and E–B
tiation from T cells on the basis of
free-flow electrophoresis, surface com-
of cholinesterase, and absence of α-
activity has already been commented
on.

lin-bearing B lymphocytes have been
phoid tissues at 11½ weeks of gestation

urface immunoglobulin stained with fluorescein-
obulins antiserum.

Table 2.2
Physicochemical Properties and Biologic Functions of Immunoglobulins

Property/function	IgG	IgM	IgA	Secretory IgA	IgD	IgE
Molecular weight	148,000	1,000,000	160,000	385,000	170,000	200,000
Sedimentation constant ($S_{20}w$)	6.9	18–150[a]	7–17[a]	11–19	6.2	8.2
Biologic half-life (days)	22	5	6	?	2.8	2.3
Synthetic rate (mg/kg/day)	34	6.7	25	?	0.4	0.03
Fractional catabolic rate (%/day)	3	14	12	?	35	89
Mean serum conc. in adult (mg%)	1150	135	280	0	1.8	0.005
% of total immunoglobulin body pool	76	9	13	?	1.5	0.04
Carbohydrate content (%)	2.9	11.8	7.5	?	12	11.6
E_{280} mu	13.8	13.3	13.4	?	13.5	15.2
Peptide chains	$\gamma_2\kappa_2$ $\gamma_2\lambda_2$	$(\mu_2\kappa_2)_5\cdot J$ $(\mu_2\lambda_2)_5\cdot J$	$\alpha_2\kappa_2$ $\alpha_2\lambda_2$	$(\alpha_2\kappa_2)_2 Sc\cdot J$ $(\alpha_2\lambda_2)_2 Sc\cdot J$	$\delta_2\kappa_2$ $\delta_2\lambda_2$	$\epsilon_2\kappa_2$ $\epsilon_2\lambda_2$
Placental transfer	+	0	0	0	0	0
Complement binding	+	+	0	?	0	0
Primary antibody response	0	++	0	0	0	0
Secondary antibody response	++	±	±	0	0	+
Reaction with rheumatoid factor	+	0	0	0	0	++
Anaphylaxis	0	0	0	0	0	+

Antibody activity	Antitoxin, antibacterial, antiviral, incomplete Rh, and immune anti-A, anti-B isoagglutinins	Antipolysaccharide and anti-gram-negative bacteria, natural iso-agglutinins, saline anti-Rh, rheumatoid factor, heterophile antibodies	Antiviral	Antiviral, prevents mucosal transfer of macromolecules	?	Reaginic antibody, defense against parasites, potentiates IgG antitoxic reactions, augmentation of phagocytosis and local inflammatory response
Miscellaneous	—	—	—	Resists proteolytic digestion	—	May activate alternate pathway of complement

a Polymer forms.
b In selective IgA deficiency.

(Lawton *et al.*, 1972). However, plasma cells with intracytoplasmic immunoglobulin are not found in healthy fetuses till about 20 weeks (van Furth *et al.*, 1965). Antigenic stimulation such as infection accelerates the developmental process, whereas a germ-free environment retards it.

IgM is the first immunoglobulin detected, as early as 10 weeks of human gestation (Gitlin and Biasucci, 1969). Soon after this time, IgA, IgD, and IgG are also found (Chandra, 1975b). Almost all of the fetal serum immunoglobulin consists of IgG which is selectively transferred from the mother across the placenta, a process in which the Fc component of the Ig molecule plays an important role. The concentration of serum IgG rises progressively with increasing gestational age and body weight (Chandra, 1975c) (Fig. 2.4). The various subclasses of IgG are acquired by the fetus at different rates, IgG_1 being transferred most effectively. Also, IgG_1 transfer is affected the most in placental insufficiency (Chandra, 1975b). Before and after birth, the concentration of immunoglobulins is determined largely by the extent and duration of antigenic stimulation. Raised cord blood IgM is highly suggestive of infection during intrauterine life (Alford *et al.*, 1975). In developing countries where infectious diseases are common and standards of sanitation are poor, there is a precocious rise in immunoglobulins to adult levels by the age of 2–5 years (Chandra and Ghai, 1972). Genetic differences also affect the development of immunoglobulins. Recent studies in mice have demonstrated the inherited interstrain variability in the quality or affinity of antibody produced (Morgan and Soothill, 1975). The role of the environment is shown by the observation that animals reared in germ-free conditions have poor and delayed development of circulating immunoglobulins.

There is a large body of data on the physicochemical and biologic properties of the major immunoglobulins classes (Table 2.2), each of which has a characteristic heavy chain and either kappa (κ) or lambda (λ) light chain (Turner and Hulme, 1971). The antigen recognition and binding mechanism is located in the portion of the immunoglobulin molecule (Fab) made up of the light chain and a portion of the heavy chain (Fd) which can be prepared by proteolytic digestion of the whole molecule. The biologic function is located in the portion of the heavy polypep-

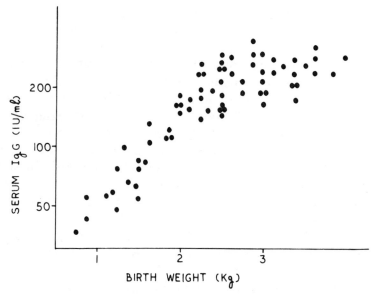

Figure 2.4. Serum IgG related to body weight. There is a significant direct correlation between the two measurements and a plateau is reached in infants who weigh 2250 g or more. From Chandra, 1975c. Copyright 1975, American Medical Association.

tide chain (Fc) split off by papain digestion. Following antigen-binding by the immunoglobulin molecule, the Fc component undergoes a conformational change to impart functional activity to it. On the basis of differences in the detailed amino acid sequence of the heavy polypeptide chain, four subclasses of IgG (G_1, G_2, G_3, G_4) and two each of IgA and IgM, and structural variants within each subclass (allotypes), are recognized.

2.2. PHAGOCYTES

Metchnikoff's pioneering observations at the turn of the century pointed to the role of leukocytes as a first-line defense barrier against bacterial invasion. Phagocytes present in the vascular compartment (neutrophils, eosinophils, basophils, monocytes) constitute a highly mobile and readily available force. The reticuloendothelial system of tissue histiocytes is constituted by

Kupffer cells in the liver, macrophages in the spleen, lymph nodes and pulmonary alveoli, and microglial cells in the central nervous system. The important role played by phagocytes in defense against bacterial and fungal disease is illustrated by chronic granulomatous disease, in which impaired killing of phagocytized intracellular bacteria results in a severe illness characterized by suppuration of the lymph nodes, liver, lungs, and other organs, often causing death in the first two decades of life.

There have been rapid recent developments in the field of phagocyte biology. These encyclopedic details are discussed in several exhaustive reviews published in the last three years (Pollard and Weihing, 1974; Stossel, 1974, 1975; Ward, 1974; Wilkinson, 1974a; Bellanti and Dayton, 1975; Berlin *et al.*, 1975; Cline, 1975; Humbert *et al.*, 1975; Klebanoff, 1975; Cheson *et al.*, 1977). This section gives a conceptual summary of some biological aspects.

The process by which polymorphonuclear leukocytes afford protection can be divided into ten stages: production, mobilization, opsonization, recognition of and attachment to the object of phagocytosis, ingestion, degranulation, metabolic activity with associated generation of microbicidal free radicals, killing, and digestion of the inactivated particle. Malfunction of any of these steps, singly or in concert, may predispose the host to frequent infections.

Granulocytes originate from the pluripotent stem cell, first noted in the human fetal liver at about 2 months of gestation, and later on in the bone marrow. The morphologic stages of differentiation include myeloblasts, promyelocytes, myelocytes, metamyelocytes, band cells, ultimately maturing into segmented polymorphonuclear leukocytes. The functional ability of the fetal phagocyte system has not been adequately studied. In the neonatal period, the reduced phagocytic capacity of neutrophils is due mainly to lack of serum factor(s) such as complement components involved in opsonization, since the defect can be almost completely corrected by employing serum of adults. The cord blood leukocytes have a distinctly reduced ability to kill intracellular bacteria and fungi.

Recent advances in cell-labeling (Thakur *et al.*, 1976) have provided a much-needed stimulus for studies on neutrophil kinetics and traffic. Experiments employing radiolabeled tritium, chromium-51, phosphorus-32, and the newcomer, gamma-emitting indium-111, showed that, after release from the bone marrow, neutrophils remain in the vascular compartment for about 6 hours, either circulating in the fast-flowing bloodstream or adhering to the endothelial lining of the vessels. The cells migrate to the tissues, drain to the lymph nodes, and reenter the circulation. The cell is attracted to the site of action by chemotactic factors released from the tissues invaded by the pathogen. Neutrophil movement depends upon the microfilaments actin and myosin, and microtubules (Stossel, 1974).

The initial step before endocytosis of particles can occur involves the coating of bacteria with antibodies and heat-labile components, mainly complement components. IgG and C3 receptors have been identified on the surface of mononuclear phagocytes (Huber and Fudenberg, 1970), but the chemical nature of such receptors is not known. The participation of C3 is essential for immune adherence of IgM-coated cells.

There is evidence to suggest that neutrophil receptor sites for the Fc portion of IgG are involved in the phagocytosis of bacteria coated with IgG antibodies (Quie *et al.*, 1968). Chemical agents which act on the free sulphydryl group, and oxidizing agents, inhibit phagocytosis by their interaction with IgG receptors, whereas reducing agents have no such effect. Proteolytic trypsin inactivates heat-labile opsonic receptors on phagocytes but has no effect on IgG receptors. The biochemical and biophysical bases of bonding between particles and cells, endocytosis, and fusion of granules with phagocytic vacuoles are not completely understood. Phagocytosis of particles is dependent on glycolysis, but oxidative consumption and oxidative phosphorylation are not essential. In addition, the presence and physicochemical character of the bacterial envelope and the autolytic enzymes released by the pathogen are important determinants in the interaction between phagocytes and microorganisms (Elsbach, 1973).

Once inside the phagocytic vacuole, a series of metabolic and enzymatic processes are activated, all of which are aimed at the ultimate destruction of the pathogen. Lysosomes, which are polymorphic packets of more than a dozen lytic enzymes, fuse with phagocytic vacuoles to form phagolysosomes. Hydrolytic enzymes, including ribonuclease, deoxyribonuclease, phosphatases, cathepsins, plycosidases, sulfatases, protease, phospholipase, elastase, mannosidase, lysozyme, glucuronidase, glucosaminidase, triglyceride lipase, and iron-chelating lactoferrin, all of which are associated with neutrophil granules or are found in cell sap, are poured into this space. Many intraleukocytic cationic proteins possess bactericidal activity, whereas other proteins have fungicidal property.

Phagocytosis is associated with a 10-fold increase in hexose monophosphate shunt activity, increased consumption of oxygen and glucose, and increased production of hydrogen peroxide. The hexose monophosphate shunt was thought to play a critical role as illustrated by metabolic and bactericidal defects in neutrophils from patients with absent glucose-6-phosphate dehydrogenase. However, it is probably a secondary event involved in the regeneration of reducing substances including glutathione and ascorbic acid, and inactivation of oxygen radicals and peroxide which diffuse out of the phagocytic vacuole into the cytoplasm. Molecular singlet oxygen, superoxide radical, hydrogen peroxide, and hydroxyl radicals are important in bacterial killing (Klebanoff, 1975). Cyanide-insensitive NADH-oxidase or glutathione reductase may provide the initial step, leading to formation of hydrogen peroxide. The continuing availability of $NADP^+$ to serve as an electron receptor for the dehydrogenases is crucial. Granulocyte enzyme myeloperoxide acts in concert with the hydrogen peroxide–halide system, and amplifies or mediates the latter in the killing of bacteria and fungi. Lysozyme (muramidase), which can hydrolyze the cell–wall mucopolypeptides of bacteria, probably provides the final blow to the microbe weakened by assaults by other bactericidal mechanisms.

2.2.1. Monocytic Phagocyte

The kinetics and functions of fixed and wandering macrophages are less well-defined than those of the polymorphonuclear leukocyte. Promonocyte in the bone marrow differentiates successively into the monocyte and then the macrophage. These cells have a rapid turnover and are affected by corticosteroids as well as cytotoxic drugs. Upon being stimulated, the macrophage becomes bigger in size and the morphological details become more complex. Several different types of intracellular granules are recognized, which play an important role in digestion of phagocytosed material. Ingestion is facilitated if the particle is opsonized by complement C3, immunoglobulins (IgG_1, IgG_3), migration inhibitory factor produced by stimulated lymphocytes, and other substances. The macrophage has cell–surface receptors for these molecules and forms rosettes with antibody coated red cells.

Activation of the macrophage increases the enzymatic potential of the cell. The microbicidal pathways include peroxide, superoxide, malonyldialdehyde derived from unsaturated fatty acids, lysozyme, lactoferrin, and interferon. Besides phagocytosis and intracellular digestion, the macrophage plays a critical role in antigen "processing" which confers heightened immunogenecity on the substance. The mechanism involved is not clear. Soluble factors as well as cell–cell contact may be important.

Pitt (1977) referred to the monocytic macrophage as "a sweeper, a mediator, and an instructor." The complex biology of the macrophage is far from completely known. The current knowledge has been reviewed recently (Carr, 1973; Wilkinson, 1974b; van Furth, 1975; Nelson, 1976).

2.3. COMPLEMENT SYSTEM

The complement system consists of a complex set of 11 distinct serum proteins which can be activated by a variety of agents such as antibodies, microbial products, and enzymes. The various complement components interact in a sequential fashion to mediate and amplify many of the biological effects of immune reactions, chemotaxis, immune adherence, opsonization, anaphylaxis, cell lysis, and bacterial inactivation. The potential ability of this system to damage the host's own cells is limited by the spontaneous decay of activated complement and control by inhibitors and destructive enzymes. The complement system also interdigitates with many other systems of the host.

In the last 10 years, considerable progress has been made in the isolation and characterization of individual components of complement and the development of immunochemical quantitation and functional assays (Muller-Eberhand, 1975; Colten, 1976; Johnston and Stroud, 1977). The genetic heterogeneity of complement proteins is now recognized. The number and variety of confusing names which evolved during the initial discoveries have now given place to a standardized nomenclature.

In man, the synthesis of complement components C1, C2, C3, C4, and probably other components, starts in early fetal life, around 8 weeks of gestation. The complement activity in fetal serum is the result of active fetal synthesis, since the placenta is an effective barrier to passage of complement either from or to the fetal circulation. The synthesis of specific complement proteins *in vitro* by isolated fetal tissues has been demonstrated. Complement components have been detected in sera of fetuses borne by genetically deficient mothers. The electrophoretic patterns of complement proteins exhibiting genetic polymorphism often differ in paired maternal-fetal samples. The principal sites of synthesis differ from one component to another: C1 by epithelial cells of the small and large intestine and the genito-urinary system except the kidney; C3, C5, C6, C9, and C1 esterase inhibitor by the hepatic parenchymal cells; and C2 and C4 by macrophages in various tissues. It appears that all the subunits of C1 are synthesized in a single cell type. The genetic loci controlling the synthesis of C3, C4, and C2 have been mapped on human chromosome 6 close to the major histocompatibility complex.

There are two major pathways of complement activation (Fig. 2.5). In the classical pathway, C1q serves as the recognition unit, C4, C2, and C3 as the activation system, and C5, C6, C7, C8, and C9 as the membrane attack unit. The three subunits of C1, namely C1q, C1r, and C1s, are held together by noncovalent bonds in the obligatory presence of Ca^{++}. A single molecule of IgM antibody on the surface of a cell is able to bind C1. In the case of IgG, two adjacent molecules are requried for such binding, an infrequent event since antibodies are scattered randomly over the cell surface and the probability of two IgG molecules' occupying adjacent sites is small. Enzymatic activation of other complement components is initiated by activated C1s. The process of enzymatic cleavage exposes a binding site with a short life on the activated complement component. The intermediate complexes are extremely unstable. The reactions

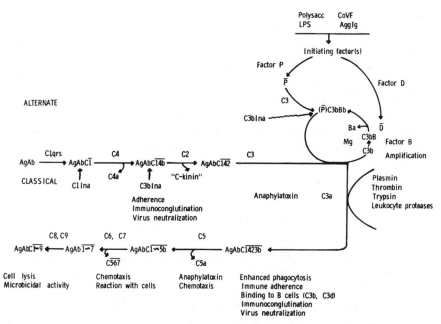

Figure 2.5. The complement cascade. Complement proteins interacting in a sequential manner generate activated forms (indicated by a bar above the symbol) to liberate biologically active fragments (designated by a lowercase letter), and eventually to cause cell lysis and death. Complement activation may be initiated in C1q (classical pathway) or through a bypass mechanism which acts on C3 (alternate pathway).

involving the late-acting complement components C7 to C9 are not yet well understood. Lesions on cell surfaces produced by complement activation can be seen in electron micrographs, and are quite uniform in size although the latter varies with each species. The internal diameter of the membrane lesions produced by human complement is 10–11 nm.

The alternate pathway bypasses C1, C4, and C2, and is activated by immunoglobulin aggregates, polysaccharides, and cobra venom factor (Fearon and Austen, 1975). It consists of the initiating factors, properdin, C3 or proactivator C3PA (Factor B, glycine rich β-glucoprotein), and C3 proactivator convertase (Factor D). It merges with the classical pathway at the C3 stage, sharing the membrane attack unit.

Alterations in the quantity and activity of complement components have been described in many clinical situations (Lachmann, 1975). Deficiencies of isolated proteins are rare and often inherited. In many acquired diseases, the complement system is an active participant in pathophysiologic processes. An increase in serum concentration is a part of an acute-phase response, whereas a decrease may be due to reduced synthesis, accelerated consumption, and fixation to tissues, or loss through the gut or the kidney.

2.4. OTHER FACTORS OF HOST RESISTANCE

A number of other mechanisms provide protection from invading pathogens. These are listed in Table 2.1, but the discussion of each of them is beyond the scope of this monograph (see Bellanti, 1970; Stiehm and Fulginiti, 1973; and Gell *et al.*, 1975).

2.5. CONCLUDING REMARKS

The principal tiers of antigen-specific and antigen-nonspecific immunity have been described separately in the above sections. It is becoming increasingly clear, however, that cellular interactions, both facilitatory and inhibitory, are the key to the generation of optimal immune response, which is further amplified by other factors, such as the complement system. Cellular and humoral phenomena can influence each other. For example, the ingestion of foreign particles by phagocytes is promoted by antibodies, complement component C3, and other serum factors. Also, antibodies activate the complement system and fix its constituent proteins on the invading cell's surface. The uniqueness of the immune system lies in the extreme heterogeneity of cell populations capable of highly specific interactions without the necessity of any physical segregation. Burnet views it as "a homeostatic and self-monitoring system whose function is to maintain the genetically defined integrity of body substance, which end it achieves by transient interchanges of information from random contact between fully mobile units'' (1976, p. 158).

Primary, often inherited, deficiencies of host defenses are associated with an increased incidence of infection of variable severity, cancer, atopy, immune complex diseases, autoimmune disorders, and reduced survival. Much more common is the secondary suppression of one or more protective mechanisms by a variety of pathological states, including infection, malignancy, therapy with corticosteroids and antimetabolites, irradiation, surgical resection, thermal injury, anesthesia, and surgery. On a global scale, however, nutritional deficiency is the commonest cause of secondary impairment of immunocompetence.

3

ASSESSMENT OF NUTRITIONAL STATUS

The definition of undernutrition has proved to be more difficult than might be expected for so prevalent a condition. The ease of clinical recognition of the two polar entities of kwashiorkor (from the Ga language of Ghana, meaning "displaced child") (Fig. 3.1) and marasmus (Greek *marasmos,* withering) (Fig. 3.2) does not belittle the semantic and diagnostic problems surrounding less severe and intermediate forms of nutritional deficiencies.

3.1. ENERGY-PROTEIN UNDERNUTRITION

The World Health Organization defines protein-calorie malnutrition as "a range of pathological conditions arising from coincident lack, in varying proportions, of protein and calories, occurring most frequently in infants and young children and most commonly associated with infection" (1973, p. 1). The world prevalence of marasmus exceeds that of kwashiorkor at least fourfold. The obsession with protein deficiency has given place to the recognition based on facts that the major nutritional deficiency in the world today is that of total food intake (calories or energy). It is logical, therefore, to use the term *energy-protein undernutrition* to provide the appropriate stress to energy deficit. Based on careful dietetic surveys, epidemiological data, and metabolic-endocrinal observations, it has been suggested that kwashiorkor and marasmus represent two aspects of nutritional

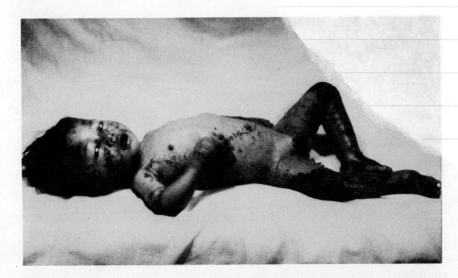

Figure 3.1. Kwashiorkor in a young Indonesian boy. The striking manifestations include extensive skin changes, edema, and mental agony. WHO photo by Dr. Liem Tjay Tie. Courtesy World Health Organization, Geneva.

imbalance and adequacy rather than two separate entities with different dietary causes.

Gomez *et al.* (1956) and Jelliffe (1966) recommended the use of weight loss as the index of severity of malnutrition. The Wellcome classification (Table 3.1) (FAO/WHO Expert Commit-

Table 3.1
Wellcome Classification of Energy–Protein Undernutrition[a]

Class	Body weight as % of standard[b]	Edema	Deficit in weight for height[c]
Underweight child	80–60	0	minimal
Nutritional dwarfing	< 60	0	minimal
Marasmus	< 60	0	++
Kwashiorkor	80–60	+	++
Marasmic kwashiorkor	< 60	+	++

[a] From FAO/WHO Expert Committee on Nutrition (1971).
[b] Reference standard is the 50th percentile for age on the Harvard growth curves.
[c] Weight for height = (weight of patient/weight of normal subject of same height) × 100.

ASSESSMENT OF NUTRITIONAL STATUS 33

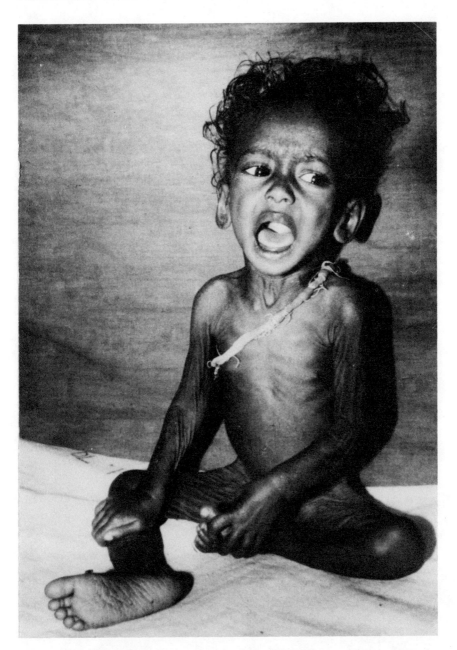

Figure 3.2. Marasmus in a 13-month old Indian girl weighing 3.94 kg. Note the gross wasting, wrinkled skin, growth failure, and misery. Courtesy World Health Organization, Geneva.

Table 3.2
Anthropometric Assessment of Nutritional Status[a]

Index	Interpretation	Advantages	Disadvantages	Observer error
Weight (% of standard)	<60% severe, 60–80% moderate, 80–90% marginal (?). Assesses current nutritional status.	Simple, reproducible.	Requires accurate age and ethnic reference standard. By itself, it does not distinguish wasting from dwarfism.	100–200 g
Length (percentile)	<3rd severe, 3rd-10th moderate, 10th-25th marginal(?). Assesses past and chronic EPU, especially in early childhood.	Simple, reproducible.	Influenced by genetic and other factors. Requires accurate age.	0.5–1.5 cm
Head circumference (percentile)	<3rd severe, 3rd-10th moderate, 10th-25th marginal(?). Assesses past and chronic EPU during fetal life and early childhood.	Simple, reproducible.	Influenced by inherent defects of brain and skull.	<0.5 cm
Skin-fold thickness	<3 mm moderately severe, 3–5 mm mild. Assesses current nutritional status and body composition.	Accurate, reproducible, simple.	Requires an expensive calliper, measurements vary with site.	1.0 mm

Measure	Interpretation	Advantages	Limitations	
Mid-arm circumference	Assesses current nutritional status.	Simple, reproducibility fair.	Reference standards not available for all groups (age, ethnic).	<0.5 cm
Weight for height age (%)	<75% severe, 75–85% moderate, 85–90% marginal (?). Assesses body build and current nutritional status.	Simple, reproducible, age-sex independent, distinguishes wasting from dwarfism.	—	—
Chest/head circumference ratio	<1 EPU	Simple, reproducibility fair, age-sex independent.	Applicable only for age group 1–2 years.	—
Mid-arm/head circumference (%)	<25% severe, 25–30% moderate. Assesses current nutritional status.	Simple, age-sex independent, reproducibility fair.	Applicable only for age group 0–4 years.	—

Note: EPU = Energy-protein undernutrition.
[a] From Chandra (1977h).

tee on Nutrition, 1971) has the merits of simplicity and uses the sensitive and reproducible measure of weight. It can be employed for valid international and longitudinal comparisons since a single set of standards is used. However, it requires knowledge of the correct age, which is not too accurate a parameter in rural areas of developing countries with varying calendar systems. The classification has been mauled also by reasoned arguments that it confuses type and severity of nutritional deficiency, and that the magnitude of deficit in weight for height cannot be quantitated by using terms such as "minimal" and "++." Waterlow and Rutishauser (1974) suggested that the duration over which energy-protein undernutrition has occurred was important and distinguished present malnutrition ("wasting," estimated by weight related to height) from the effect of previous deficits ("stunting," measured by height for age).

The problems of an all-encompassing classification and the interpretation, advantages, and disadvantages of various indices of nutritional status have been recently reviewed (Chandra, 1977h). Progressive depletion in nutritional status results, successively, in biochemical and physiologic adaptation, changes in the size of the body and its component parts, altered concentration of tissue constituents, biochemical lesions, disordered physicochemical organ function, histomorphologic changes in tissues, and, finally, clinical manifestations (Chandra, 1977h). Studies which include data based on the relatively late indices obviously cannot be compared with others that examine earlier subclinical stages of nutritional deficiency.

Anthropometric measurement is the most extensively employed index of nutritional deficiency, past and present. It has the advantages of being simple, inexpensive, reproducible, quantitative, and accurate. Dietary intake and assimilation are the most relevant of all environmental factors which interact with genetic potential to determine physical size. A summary of anthropometric measurements currently in common use is given in Table 3.2. A ratio of two indices obviates the necessity of an accurate assessment of age. Repeated measurements estimate growth velocity and are more useful than a single-point analysis.

The accurate assessment of various grades of nutritional deficiency requires biochemical measurements. Biochemical estimations can be carried out on a number of body tissues, for example, muscle, liver, bone. In practice, tests are limited to the two easily accessible body fluids, blood and urine (Table 3.3). Appropriate tests must be selected to serve the requirements of the particular study. The ideal test requires that the sample should be easily collectable (random sample of urine, finger-prick blood) and stable during transport, not influenced by recent dietary intake, cheap, simple, sensitive, specific, and reproducible.

Table 3.3
Laboratory Tests for Assessment of Nutritional Status

Nutritional deficiency	Test	
	Blood	Urine
Energy–protein	Transferrin (S)	
	Albumin (S)	
	Urea (B, S)	Urea[a]
	Amino acids[b] (S)	
		Creatinine
	Amylase	
Vitamin A	Vitamin A (S)	
	Cartene (S)	
Vitamin B_1	Transketolase (R)	
		Thiamine[c]
	Pyruvate and lactate (B)	
Vitamin B_2	Riboflavin (R)	Riboflavin[c]
Vitamin B_6		Xanthourenic acid following tryptophan load
Niacin		N-methylnicotinamide
		N-methyl-2-pyridone-5-carbonamide
Vitamine B_{12}	Vitamin B_{12} (S)	
	Hemoglobin (B)	
	Morphology (R, P)	
Folic acid	Folate (S, R)	
	Hemoglobin (B)	
	Morphology (R, P)	
Iron	Transferrin saturation (S)	
	Iron (S)	
	Hemoglobin (B)	
	Morphology (R)	
Iodine	T4, T3, TSH	Iodine

Note: (S) = serum; (B) = blood; (R) = red cells; (P) = polymorphonuclear leukocytes.
[a] Expressed as a ratio of creatinine.
[b] Ratio of essential/nonessential or valine/glycine provides a fair estimate.
[c] Random sample as well as after "loading."

3.2. VITAMIN AND MINERAL DEFICIENCIES

Deficits of vitamins, minerals, and trace metals singly, or more often in combination, are invariably encountered in individuals with energy–protein undernutrition. Once again, the diagnosis of overt syndromes is not difficult. For example, vitamin A deficiency is recognized if night blindness, xerophthalmia, and corneal opacity are observed. However, mild and subclinical deficits are not easy to pick up or quantitate but are likely to be functionally important, as has been brought home to us by the example of latent iron deficiency (Chandra, 1973a, 1975d, 1976a, 1976d, 1977b; MacDougall *et al.*, 1975).

3.3. FETAL MALNUTRITION

The definitive diagnosis of fetal malnutrition is even more difficult than that of postnatal malnutrition. A variety of etiopathogenetic factors lead to intrauterine growth retardation represented by infants who are small-for-gestational-age. An accurate assessment of birth weight and of gestation by maternal history of the last menstrual period, and neurodevelopmental examination of the infant, is required for the neonatal diagnosis of fetal malnutrition. However, the diagnosis by purely clinical means is unreliable except in those with gross deficits of growth. In autopsy materials, Anderson (1972) showed that the ratio of brain weight to liver weight remained relatively constant throughout fetal development (mean, 2.8; range, 1.7–4.1). Infants with body weight less than 1 SD below the mean body weight for gestation had an elevated brain/liver-weight ratio of 4.5 or more. Survival time had no significant effect on the index. He suggested that this ratio may be employed as an index of prenatal nutrition.

Antenatal diagnosis of intrauterine growth retardation has been attempted by ultrasound examination of the fetus (Campbell, 1970) and measurement of specific protein and steroid products of the fetoplacental unit in maternal serum and urine (Gruenwald, 1975; Klopper, 1976). Urinary or plasma estrogens, human placental lactogen, human chorionic gonadotrophin, and heat-stable alkaline phosphatase are the most widely used biochemical indices. There is a good correlation between serum concentrations of placenta-specific materials and fetal weight, which may be expected based on the known dependence of fetal size on placental function. A recent preliminary report suggests that measurement of pregnancy-specific β_1-glycoprotein in maternal blood may provide a sensitive, easy, and practical index of fetal well-being and growth (Gordon et al., 1977).

Another approach to the antenatal assessment of fetal nutrition focused on maternal leukocyte enzymes and metabolites (Metcoff, 1974). Nutritional deficiency alters the energy metabolism of cells (Yoshida et al., 1967). Leukocytes of infants with marasmus and kwashiorkor contain reduced amounts of the metabolites oxalacetate, pyruvate, lactate, and adenine nucleotides. Activities of mitochondrial adenylate kinase and cytoplasmic pyruvate kinase are reduced. Similar data were obtained on the cord-blood leukocytes of infants born small-for-gestational-age and in the leukocytes of their mothers, indicating that the metabolism of the maternal leukocyte might reflect fetal nutrition and development. Birth weight of the infants correlated directly with RNA polymerase activity of leukocytes from the peripheral blood of their mothers. This was in contrast to the findings in the placentas. The mechanism underlying such changes in enzymes is not clear. Several interacting variables may influence enzyme activity, including a change in the number of enzyme molecules, alterations in intracellular modulators, and change from one molecular form to another. Mameesh et al. (1976) have recently looked at the kinetic properties of pyruvate kinase in the mothers of infants with fetal malnutrition. Pyruvate kinase, one of the rate-limiting regulatory enzymes in glycolysis, exists in two major isoenzyme forms in mammalian tissues: type L in the liver, parenchymal cells, and erythrocytes; and type M in muscle and leukocytes. Based on Michaelis–Menten kinetics, the M isoenzyme is considered nonallosteric whereas the Type L isoenzyme is modified by phosphoenolpyruvate, adenosine triphosphate, L-alanine, K^+, and fructose-1,6-diphosphate. The activity of leukocyte pyruvate kinase from mothers delivering small-for-gestational-age infants was reduced with respect to the substrate phosphoenolpyruvate, but was less responsive to activation by fructose-1,6-diphosphate, irrespective of the presence or absence of L-alanine. This and other data suggested that pyruvate kinase in maternal leukocytes during pregnancy is influenced by the same modulators that modify the activity of the L-type enzyme. It is possible that ethnic and geographic factors as well as nutritional deficiency or imbalance may affect the leukocyte metabolism. Since the human fetus depends upon glycolysis for its energy needs, impaired enzyme activity in intrauterine growth retardation could lead to a reduction in available metabolic energy required by rapidly proliferating embryonal tissues. The metabolism of maternal leukocytes including the kinetic response of pyruvate kinase to allosteric modulators during pregnancy may provide an additional test for the antenatal diagnosis of fetal malnutrition.

3.4. CONCLUDING REMARKS

The problems of the definition and grading of deficiencies of the many nutrients consumed by man make the interpretation of published literature on nutrition-immunity-infection interactions difficult. Many reports on this topic have not clearly defined the nutritional status of the subjects that were evaluated. This is particularly true of human studies.

4

INFECTIONS IN UNDERNOURISHED INDIVIDUALS

The clinical impression that nutritional deficiencies generally reduce the capacity of the host to resist infection and its consequences is widely accepted. Published evidence which suggests that nutritional status conditions the individual to infectious disease is based on several different types of data: higher point prevalence of infection in undernourished subjects attending the clinic or hospital or surveyed in the community, more frequent and more severe complications following an infectious illness in the malnourished, higher infection-related mortality rate in malnourished populations, frequent presence of complicating infection in children dying of kwashiorkor or marasmus, reduction in infection rate associated with improvement in dietary intake, higher rate of occurrence of infections during war or blockade or following famine. However, many of the studies suffer from one or more faults in sampling: lack of inclusion of a sufficient number of subjects with different severity of nutritional deficiency; retrospective analysis; inadequate diagnostic workup for nutritional status and for the presence and etiology of infection; lack of consideration of other environmental determinants, for example sanitation and personal hygiene; inadequate statistical analysis; etc.

Scrimshaw *et al.* (1968) reviewed all the studies to date on the effect of malnutrition on resistance to infection and summarized that multiple nutritional deficiencies increase the incidence and/or severity of tuberculosis, acute diarrheal disease, rickettsia, infectious hepatitis, measles, amebiasis, and respiratory disease. The extensive epidemiologic data from field studies in Guatemala summarized by Scrimshaw (1970) and Mata (1975) suggest synergistic interactions of malnutrition and infection. This applied also to small-for-gestational-age, low-birth-weight infants who showed greater rates of infection with *Shigella, E. histolytica, G. lamblia,* and *Candida*. Dysentery and bronchopneumonia were more frequent in children in the lower quartile of growth than in those in the upper quartile (Mata *et al.*, 1971). In Australian aboriginal children, Jose and Welch (1970) found that intestinal parasites reached heavier loads in growth-retarded subjects but played no part in initiating the syndrome. A high proportion of children with deafness, or dying from gastroenteritis and/or pneumonia, had a previous history of nutrition-related growth failure.

Of children with moderate-severe malnutrition admitted to hospitals in Asia and Africa, 30 to 65% have active infection, and there is a high sepsis-related mortality in these patients (Smythe and Campbell, 1959; Phillips and Wharton, 1968). Malnourished children show a tendency to develop gram-negative septicemia, disseminated herpes simplex infection, afebrile or anergic response to infection, and gangrene rather than suppuration.

A number of studies have pointed to the high mortality from measles in the developing countries, which may be attributable in part to the poor nutritional status of the population (Ghosh and Dhatt, 1961; Morley, 1962, 1964; Taneja *et al.*, 1962; Gordon *et al.*, 1965). Malnourished children suffering from measles may not show a rash. There was a high frequency of giant-cell pneumonia, and thymolymphatic atrophy was less marked than in the children dying with energy–protein undernutrition who did not have measles (Smythe *et al.*, 1971).

Hepatitis-associated antigen was found in a higher proportion of Indian and Thai patients with energy–protein undernutrition than in the general population (Suskind *et al.*, 1973; Chandra, 1977i). The incidence of HBsAg was approximately 3 to 8 times higher than in age–sex-matched well-nourished controls from the same community. The higher HBsAg detection rate in marasmus compared with the incidence in kwashiorkor may be explained on the basis of a longer time period over which an opportunity for exposure and infection existed in the former. Information on the prevalence rate of other hepatitis antigens and subtypes is not available. The frequency of detection of HBsAb was slightly lower, though statistically it was not significant (Chandra, 1977i). This may well be due to depressed cell-mediated immunity and helper function necessary for mounting an adequate antibody response to viruses, as also seen in Indian childhood cirrhosis (Chandra, 1975i, 1976f). Cellular immunity is consistently impaired in nutritional deficiency. Other syndromes with the common denominators of depressed CMI and susceptibility to develop antigenemia with HBsAg include lepromatous leprosy, Down's syndrome, and leukemia–lymphoma.

Generalized, often fatal, infection with herpes simplex virus is rare, apart from the perinatal period. In Africa, it has been observed frequently in association with kwashiorkor in children 6–24 months of age. Templeton (1970) reported five such cases characterized by necrotic hemorrhagic lesions in various organs, especially under the capsule. The diagnosis was confirmed on culture, locating inclusion bodies in liver cells, and by indirect immunofluorescence. The adrenal cortex was affected exclusively and the medulla was always normal. In the liver, the surface and cut sections showed characteristic whitish spheres each surrounded by a "halo" of hemorrhage, which stood out in the fatty background. There was a central zone of complete necrosis surrounded by virus-infected cells. The absence of inflammatory cells was striking. Local interferon production has been shown to be important in the healing of experimental herpes infection in mice, and this, together with depressed cell-mediated immunity, may be a contributory factor in the spread of herpesvirus infection.

Morehead et al. (1974) reviewed 35 consecutive admissions of Thai children with moderate-severe malnutrition and found 32 to be having one or more infections. Forty "major" infections, defined as those potentially life-threatening, were present in 24 patients, more commonly in those with edema and with clinical evidence of vitamin A deficiency. Infections included pneumonia, genito-urinary infection, septicemia, wound infection, and isolation of enteropathogens. Septicemia was associated regularly with edema and low serum proteins. Significant bacteriuria was often found in the absence of pyuria. In some children, more than one major and/or minor infections were observed. Thirty-seven "minor" infections (otitis media, otitis externa, skin infections, conjunctivitis) were found in 26 patients. A variety of pathogenic organisms was isolated, the same organism being found in more than one site in some patients. Mixed infections were found in 44% of all positive cultures, in contrast to an incidence of 5-10% in the well-nourished. Specific pathogens isolated included gram-positive cocci in 26 instances, gram-negative bacteria in 41, *Shigella sonnei* in 3, *Salmonella* in 2, Mimae/Moraxella, Bacteroides, and *Mycobacterium tuberculosis* in one each. *Candida* was isolated only once, from a case of otitis externa. Pathogens generally showed marked *in vitro* resistance to multiple antibiotics.

Autopsy studies have confirmed infection to be a significant causative factor in deaths of malnourished children. Twenty-two of 25 African malnourished children on whom autopsy data were reviewed by Purtilo and Connor (1975) died of fulminant infections, including disseminated varicella, staphylococcal infection, miliary tuberculosis, cerebral malaria, measles, herpes simplex, diphtheria, and *Pneumocystis carinii*, and infestation with *Strongyloides stercoralis, Necator americanus* and *Ascaris lumbricoides*.

Hughes *et al.* (1974) examined the role of protein-calorie malnutrition as a host determinant for *Pneumocystis carinii* pneumonitis. Mean body weights and serum protein values were below normal in cancer patients with *P. carinii* infection than in matched controls without pneumonitis. Similar findings have been reported in patients with primary immunodeficiency disorders, organ transplants, and other serious underlying diseases. By the use of methenamine silver nitrate impregnation technique, the organism was found in 3 (7.7%) of 39 children with kwashiorkor and pulmonary infection, but in none of 21 well-nourished children of whom 16 had died with pneumonia. Other reports mention the occurrence of pneumocystosis in marasmic infants often wasted from chronic diarrhea (Dutz, 1970). The incidence of *P. carinii* infestation in undernutrition (Bwibo and Owor, 1970; Hughes *et al.*, 1974) is similar to its presence in children with lymphoproliferative malignant neoplasms (Hughes *et al.*, 1973).

The effect of deficiency of individual nutrients on host resistance to infection has been extensively studied in laboratory animals but such data is almost impossible to obtain in man since human malnutrition is generally the result of deficits of multiple nutrients. The reported increase in the incidence and severity of infectious illnesses in children with overt signs of vitamin deficiency, for example hypovitaminosis A and D (Clausen, 1935) is likely to reflect a more severe form of energy–protein undernutrition as well. Bean and Hodges (1954) reported a higher frequency of upper respiratory infections during a 35-day period of experimental pantothenic acid deficiency in four subjects.

Clinical surveys estimating the incidence of infection in relation to the individual's status of iron have yielded results, varying from increased to reduced prevalence. Infection, recurrent or severe, is the most common symptom for which iron-deficient children seek medical advice. MacKay (1928) observed a modest decrease in the number of episodes of bronchitis and gastroenteritis in iron-supplemented infants from low-income families in London. More recently, Andelman and Sered (1966) found that respiratory infections were significantly less in infants who were given an iron-fortified milk formula. Iron deficiency and impaired cellular immunity are common findings in patients with chronic mucocutaneous candidiasis (Higgs and Wells, 1973; Fletcher *et al.*, 1975). The skin lesions as well as immunological abnormalities reversed rapidly on administration of iron. Basta and Churchill (1974) reported a higher prevalence of acute and chronic infections in the iron-deficient anemic workers in Indonesia compared with nonanemic controls. Adults with recurrent herpesvirus infection show a higher frequency of iron deficiency than matched controls. Other studies failed to show a relationship between iron deficiency and infection frequency (Howell, 1971; Burman, 1972). A Tanzanian study (Masawe *et al.*, 1974) of older children and adults with nutritional anemia found that iron-deficient individuals had a lower frequency of infection than did patients with other types of severe anemia, but there was no evaluation of any control group of healthy subjects.

Acute diarrheal disease in developing countries is most commonly seen in young infants around the time of weaning from the breast. The syndrome of "weanling diarrhea" includes a minority of infectious gastroenteritis caused by a specific pathogen, a proportion of undifferentiated, presumably infectious, illnesses, and a bulk of nonmicrobial entities. Observations made during epidemiologic surveys in India and Guatemala suggest that nutritional deficiency is an important etiopathogenetic factor. In a study of rural preschool children around Delhi, the incidence of diarrheal illness in the undernourished group was 2–3-fold greater than in the well-nourished group (Ghai and Jaiswal, 1970). Deterioration in nutritional status due to dietary inadequacy was associated with higher spell frequency of diarrhea, whereas improvement in nutrition reduced it. In a large majority of patients with frequent loose stools, no known bacterial enteropathogen or enterovirus was isolated. The frequency of detection of enteropathogens was almost equal in patients and symptomatic controls. Quantitative and qualitative changes in the gut microflora of malnourished children (Heyworth and Brown, 1975) may be important in the pathogenesis of diarrhea and may worsen the nutritional deficiency. Structural and functional changes in the small intestine (Amin *et al.*, 1969), altered bile acid formation (Schneider and Viteri, 1974), and pancreatic atrophy may also contribute to malabsorption.

The determinants of asymptomatic contact and colonization differ from those of infection and disease. It is widely accepted that nutritional status is a critical modulating factor that influences susceptibility to infectious disease (Scrimshaw *et al.*, 1968; Chandra, 1976b). Deficiencies of individual nutrients may alter infection-related morbidity. In an undernourished person, the risk of developing infectious illness and its complications, including death, will be determined, among other factors, by the magnitude and nature of nutrient imbalance.

5

NUTRITIONAL, METABOLIC, AND IMMUNOLOGIC EFFECTS OF INFECTION

Infectious illness is regularly associated with alterations in the metabolic balance, hormonal milieu, nutritional status, and host resistance. The magnitude of these changes is related to the severity and duration of infection, the type of pathogen, the tissues involved, and the preinfection status of nutrition and immunocompetence of the individual. In some, careful serial analyses may be required to detect changes in nutrition and immune capacity. In others, overt clinical manifestations and complications point to infection-associated disturbances in homeostasis.

5.1. NUTRITIONAL CONSEQUENCES OF INFECTION

Infectious disease is invariably associated with loss of body constituents, rapid utilization of body stores of nutrients, and redistribution between various physiologic metabolic compartments. Most infections cause anorexia and a decrease in food intake. In some diseases, such as measles and herpes, soreness of the mouth in young infants may lead to inability to suck and failure of feeding from the breast or the bottle. Vomiting and diarrhea are frequent symptoms of infection and aggravate nutritional losses and malabsorption (Fig. 1.1).

Systemic bacterial infections impair xylose absorption (Cook, 1972). In a study of Nigerian infants with acute measles enteritis, the mean absolute loss of albumin measured by fecal clearance of Fe^{59}-labeled dextran was 1.7 g/day, the equivalent of about 20% of the child's normal protein intake (Dossetor and Whittle, 1975). Serum albumin falls during an attack of measles (Fig. 5.1) and it may precipitate edema.

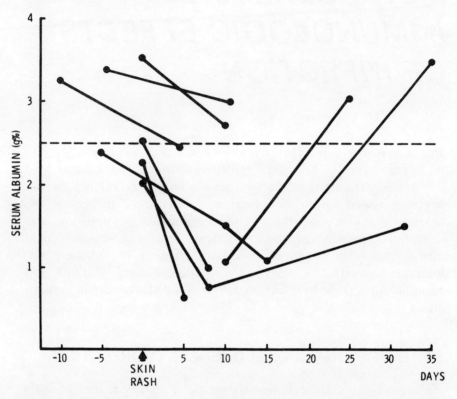

Figure 5.1. Effect of measles on nutritional status measured by serum albumin concentration. In all patients, a significant drop in albumin level occurred. In those with low-normal levels before the infection, there was a more pronounced fall in serum albumin concentration, often associated with the appearance of dependent edema. Recovery occurred several weeks following medical nutritional management.

Table 5.1
Nutrient Consequences of Infection[a]

I. Absolute losses
  Increased urinary nitrogen
  Loss of electrolytes, minerals, and proteins in vomiting and diarrhea
  Proteinuria
  Negative metabolic balance of cations, minerals, and trace elements
II. Functional wastage
  A. Overutilization
    Increased usage of metabolic substrates
    Depletion of glycogen stores
    Diversion of amino acids for gluconeogenesis
    Mobilization of fat from depots
    Increased synthesis of cholesterol and triglyceride
  B. Diversion
    Hepatic uptake of plasma nutrients, e.g., amino acids
    Synthesis of acute phase reactant proteins
    Increased hepatocytic synthesis of enzymes
  C. Sequestration
    Uptake of minerals (Fe, Zn) into parenchymal liver cells and phagocytes
    Uptake of trace elements into liver and other organs

[a] Adapted from Beisel (1972, 1975). This topic was extensively reviewed in a workshop on Impact of Infection on Nutritional Status of the Host, May 11–13, 1976, Warrenton, Virginia (published in *Am. J. Clin. Nutr.* August, 1977).

The pattern and severity of nutritional changes varies with different types of infections or inflammatory processes, and are proportionate to the severity and duration of illness. Beisel (1972, 1975) has classified nutritional wastage during infection as absolute or functional (Table 5.1). Catabolic and anabolic processes take place simultaneously, both sets contributing to depletion of nutrients. The nutritional losses are of little consequence to a well-nourished individual. In subjects with marginal or moderately severe nutritional deficiencies, however, infection may tip the balance toward overt malnutrition, such as kwashiorkor.

Serum levels of nutrients have been sequentially studied in a variety of natural and experimentally induced infections. It must be made clear that changes in concentration of a substance represent the net balance of rates of entry and egress of that constituent from the compartment, as of course of any alterations in pool size.

5.1.1. Proteins

Free amino acids have been utilized as an index of protein status. Infection is associated with an early decrease in plasma concentration of amino acids (Wannemacher, 1977), a consequence of a rapid uptake by hepatocytes engaged in synthesis of various acute-phase reactant glycoproteins. This synthetic activity is associated with sequential changes in nucleic acid metabolism and consequent production of proteins in rough endoplasmic reticulum. The hepatic flux of amino acids contributes also to gluconeogenesis. Deamination makes the carbon skeleton available for synthesis of glucose whereas the nitrogen groups give rise to increased production and urinary output of urea. Amino acids are used up not only from the free plasma pool but also from accelerated degradation of proteins in various tissues, especially skeletal muscle. It is possible that some of the released amino acids are reutilized within the same or adjacent muscle cells and only the nonutilizable excess overflows into the blood vascular compartment. The amino acids involved in these metabolic changes include alanine, glutamine, valine, leucine, and isoleucine. 3-Methylhistidine, an analogue of the amino acid histidine, is produced from the breakdown of the contractile proteins actin and myosin, and is an excellent marker to monitor the rate of muscle catabolism (Wannemacher *et al.*, 1974). 3-Methylhistidine is unique in being neither reutilized nor degraded to carbon dioxide.

In contrast to other amino acids, including tyrosine, plasma phenylalanine concentration is increased during infection associated with fever. The ratio of plasma phenylalanine to plasma tysosine is increased. Phenylalanine is poured out of catabolized skeletal muscle. This exceeds by far the slight increase in hepatic uptake of phenylalanine. It is possible that the activity of hepatic parenchymal cell enzymes phenylalanine hydroxylase (E.C.1.14.16.1) and dihydropterine reductase (1.1.99.7) is impaired, thereby contributing to reduced levels of tyrosine and increased accumulation of phenylalanine.

In most infections, most particularly in typhoid fever, there is a rapid utilization of tryptophan by the liver cells as well as via kynurenine pathways leading to the production of serotinin (Rapoport and Beisel, 1971). Purine metabolism is accelerated as a direct result of rapid turnover of body cells. This is reflected in increased renal excretion of uric acid (Nessan et al., 1974).

A consistent increase in hepatic synthesis of acute-phase reactant glycoproteins is characteristic of most infections. These proteins include C-reactive fibrinogen, alpha$_1$-antitrypsin and alpha$_1$-acid glycoprotein. This response occurs even in malnourished individuals, attesting to its fundamental importance and teleologic value in terms of survival. In addition, amino acids made available from the breakdown of peripheral tissues are required for lymphocyte and granulocyte proliferation, immunoglobulin production, and synthesis of antigen-nonspecific host resistance factors such as complement components, kinin precursors, and proteins of the coagulation system. There is a selective increase in tryptophan oxygenase (1.13.1.12) and tyrosine transaminase (2.6.1.5) (Rapaport et al., 1968). Other hepatic enzymes are probably expendable for a short time and are usually reduced during infection (Canonico et al., 1975). These include catalase (1.11.1.6), glucose-6-phosphatase (3.1.3.9), 5' nucleotidase (3.1.3.5), and urate oxidase (1.7.3.3).

5.1.2. Carbohydrates

A number of alterations in carbohydrate nutrition occur during infective illness (Rayfield et al., 1973; Rocha et al., 1973). Gluconeogenesis is enhanced. Insulin requirements rise and glucose tolerance is impaired even in a nondiabetic individual. Serum glucagon concentration is high. The accelerated output of adrenocortical and growth hormones leads to glycogen breakdown and fasting hyperglycemia, a logical metabolic response to increased energy demands.

On the other hand, fulminant endotoxemia and viral hepatitis damage may inhibit gluconeogenesis, resulting in hypoglycemia. If glycogen stores are severely depleted and the peripheral muscle mass is reduced, as in marasmus or in the small-for-gestation, low-birth-weight infant, symptomatic hypoglycemia may complicate sepsis (Yeung, 1970).

5.1.3. Lipids

In many infections, especially gram-negative septicemia, total serum lipid concentration increases. Changes in individual fractions are variable and inconsistent. The ability of the liver to synthesize ketones is impaired (Neufeld *et al.*, 1976). Early in infection, fatty acid levels increase and triglycerides and cholesterol are formed rapidly. This, together with impaired ability of peripheral tissues to remove circulating triglycerides and reduced postheparin activity of lipolytic enzymes, results in elevated serum lipid levels and fatty metamorphosis of hepatocytes. During starvation due to causes other than infection, fatty acids from fat depots are employed as a substrate for ketone formation, thereby reducing the use of amino acids for gluconeogenesis or ketogenesis. In some infected individuals, there may be a failure to initiate homeostatic mechanisms, such as accelerated mobilization of body fat, normally activated to compensate for a reduced intake.

5.1.4. Minerals and Electrolytes

During infective illness, most of the intracellular minerals, for example potassium, magnesium, zinc, sulphur, and phosphorus, are lost in proportion to the loss of body nitrogen. The precipitous fall in serum iron concentration, often observed during the incubation period, probably reflects a redistribution between different body compartments. This may be a protective step, sequestering the mineral from siderophores of the microorganism, thereby impairing its proliferative capability. The decrease in serum iron is greater than can be explained by changes in transferrin levels. A similar change in zinc concentration occurs. Zinc may contribute to the formation and stability of metalloenzymes and cell membranes or act as a cofactor in protein and nucleic acid synthesis. As opposed to iron and zinc, serum copper rises in proportion to the synthesis and release of Cu-binding ceruloplasmin.

The principal extracellular ions, sodium and chloride, may also be lost through excessive sweating, vomiting, or diarrhea. At the same time, fever stimulates the secretion of salt-retaining mineralocorticoids. Urinary sodium and chloride are markedly reduced. In fulminant septicemia, toxic damage to cell membranes may permit sodium to accumulate intracellularly. Infections of the central nervous system are often associated with an inappropriate secretion of antidiuretic hormone and dilutional hyponatremia.

5.1.5. Vitamins

The requirements and losses of the principal vitamins increase during infections and infestations. Intestinal helminths and protozoa add the factors of blood loss as well as malabsorption. Nutritional anemias due to deficiency of iron, folates, and vitamin B_{12} are almost invariably seen in patients with hookworm and *Ascaris* disease. Infections may precipitate overt manifestations of vitamin deficiency, such as xerophthalmia, pellagrous dermatitis, beriberi, angular stomatitis, etc. The ophthalmic complications of vitamin A deficiency are believed to be significantly aggravated by associated local infections. In experimental models, corneal damage from hypovitaminosis A can be prevented or markedly delayed by maintaining a sterile environment in the eye.

5.2. METABOLIC AND HORMONAL RESPONSES DURING INFECTION

Infective illness is associated with a complex but predictable set of metabolic and hormonal changes. These alterations in the homeostatic milieu contribute to the clinical manifestations of infection. Sequential studies have delineated a stereotyped pattern of biochemical, hormonal, and physiologic responses. In addition, infection with some organisms induces pathognomonic changes in cell metabolism leading to symptoms characteristic of the causative infectious agent. Inflammatory processes mounted by the host contribute to additional metabolic alterations.

Fever is a common accompaniment of infection. The febrile response itself produces many biochemical, metabolic, and physiologic alterations. The basal metabolic rate is increased, pH rises, plasma CO_2 concentration falls, and there is hyperpnea. These alterations lead to respiratory alkalosis. If this persists, and with the increased production of acid metabolites, it may give way to metabolic acidosis. Fever promotes tissue losses of nitrogen. The urinary content of urea, α-amino nitrogen, creatinine, and uric acid are all increased. Hormonal changes during febrile response add to the metabolic alterations. Thyroid hormones show a biphasic pattern. These hormones are rapidly deiodinated resulting in a slight fall in serum protein-bound iodine, followed by an increase induced by a delayed secretion of thyroid-stimulating hormone. Changes in the production, release, and plasma levels of glucocorticoids, growth hormone, glucagon, and insulin are invariably seen.

Infection stimulates a number of anabolic responses which are intimately linked with host defense. In the bone marrow and lymphoid organs, the production of lymphocytes and phagocytes is increased severalfold. Cellular metabolism is profoundly altered. Direct oxidative metabolism of glucose and oxygen consumption are enhanced in order to generate microbicidal radicals, glycolysis is stimulated to supply energy for chemotaxis and phagocytosis, and lipid turnover is increased to regenerate membrane fragments internalized to form phagocytic vacuoles.

Two hormone-like substances produced by leukocytes and released during infectious processes have profound effects on the metabolism of other cells. The first, leukocyte endogenous mediator, is a heat-labile, trypsin- and pronase-sensitive protein of small molecular weight of about 25,000 daltons. It is unaffected by changes in pH, or by lipase and ribonuclease. The leukocyte endogenous mediator initiates the accelerated flux of amino acids and minerals from plasma to liver, and increases the synthesis of inducible enzymes and acute-phase reactant glycoproteins in hepatocytes, with compensatory decrease in the production of relatively nonessential albumin. Iron is taken up by liver cells where it combines with apoferritin to form ferritin and hemosiderin. The initial metabolic alterations are achieved without activation of the cyclic AMP adenyl cyclase system, but subsequent protein production requires new RNA synthesis. The second leukocyte mediator, endogenous pyrogen, is also of small molecular weight and acts on the hypothalamic neurones to alter the thermostatic arc resulting in fever. The physiologic role of febrile response during infectious illness is not fully understood.

In organ-localizing infections, additional losses may be superimposed on the basic host responses outlined earlier. The most significant are nutrient wastage via diarrheal stools which contain large amounts of water, nitrogen, sodium, potassium, chloride, and bicarbonate. Renal infections are often associated with loss of renal tubular function and, if severe, of glomerular function. This may end in uremia and fatal renal failure. Hepatocytic injury interferes with gluconeogenesis and may be complicated by hypoglycemia (Felig *et al.*, 1970). The ability to produce and handle bile salts and bile pigments is also affected. Ferritin is released and, unlike other infections, serum iron concentration is increased. Infections of the central nervous system are sometimes associated with an inappropriate excessive secretion of pituitary antidiuretic hormone resulting in water intoxication. In muscular wasting associated with primary muscular disorders or paralytic diseases secondary to neurologic deficit, intracellular constituents are lost in excess.

The microbe-specific metabolic responses are epitomized by the effect of *Vibrio cholerae* on epithelial cells of intestinal mucosa. The organism produces a toxin of a large molecular weight of approximately 82,000 (Sharp, 1973). The toxin binds to the receptor on the brush border of the gut epithelium and stimulates the adenyl cyclase cyclic AMP system. There is a rapid outpouring of water and electrolytes, an energy-dependent process (Keusch *et al.*, 1971). The profuse diarrhea and the resultant dehydration, electrolyte losses, and acid-base disturbance produce the characteristic clinical manifestations, complications, and mortality associated with cholera.

Interestingly, many of the metabolic and hormonal changes are initiated during the symptom-free incubation period well before any clinical manifestations are seen. The extent of metabolic changes induced by infection is directly related to the severity and duration of the illness. The higher the febrile response, the greater is the nutritional wastage. If the disease becomes chronic, the stress hormonal and metabolic changes diminish, though the cumulative losses tend to become worse, progressively encroaching upon all body tissues, including the fat, ultimately ending in cachexia. The homeostatic equilibria are set anew, with minimal additional losses being permitted.

5.3. IMMUNE RESPONSES IN INFECTION

Microorganisms entering the human body elicit four basic types of immunologic reactions producing tissue damage and disease (Coombs and Gell, 1975). Similarly, immunologic effector phenomena involved in rendering a host immune from an infection have been grouped into four major categories (Coombs and Smith, 1975). First, serum antibody reacts with microbial antigen(s) with or without other soluble molecular cofactors. Here, antibody may act as an antitoxin, neutralize virus, inhibit important enzyme systems or metabolic pathways of the pathogen, react with soluble nontoxic microbial products, or inactivate organisms by complement activation. Second, serum antibody may act together with phagocytes. Third, antibody may passively sensitize macrophages and mast cells. Fourth, the mechanism of delayed hypersensitivity may be mediated by T lymphocytes and their soluble products. More than one mode may be active in the host response to one pathogen.

Besides eliciting responses aimed at eliminating the organism and achieving protection from reinfection, infections are often associated with a transient and variable suppression of various aspects of immunologic reactivity. The characteristics of the pathogen, severity of the disease, nutritional status, and metabolic and hormonal changes are important determinants in this effect. Virtually all patients with fulminant generalized infections have suboptimal immune responses to unrelated antigens. Immunosuppression is a frequent accompaniment of infection but sometimes it is so abbreviated in duration or magnitude as to remain undetectable. Recent advances in immunological methodology have provided us with the tools with which the mechanism(s) of immunodepression during and following episodes of infection can be analyzed.

5.3.1. Cell-Mediated Immunity

Measles, miliary tuberculosis, and lepromatous leprosy are outstanding examples of infection-associated anergy. Clemens von Pirquet (1908) defined anergy as lack of reactivity or of production of "ergins" released by antigen–antibody interactions. He was the first investigator to report depressed cutaneous delayed hypersensitivity to tuberculin in patients with measles. Infectious processes implicated in the development of anergy include measles, infectious mononucleosis, rubella, mumps, varicella, influenza, yellow fever, tuberculosis, leprosy, syphilis, brucellosis, scarlet fever, histoplasmosis, blastomycosis, coccidiodomycosis, toxoplasmosis, malaria, and schistosomiasis. Andersen *et al.* (1976) conducted sequential studies of lymphocyte transformation responses in acute bacterial meningitis. Lymphocyte proliferation was depressed during the acute phases of illness, and the responses to microbial antigens were more affected than the responses to mitogens. Impairment of tuberculin reaction has also been reported after immunization with live attenuated virus vaccines (Brody *et al.*, 1964). In addition, pyogenic infections and noninfectious causes of leukocytosis may be associated with cutaneous anergy (Heiss and Palmer, 1974), with or without an associated defect of inflammatory response elicited by chemical irritant oils.

In children suffering from measles, skin reactivity to tuberculin and other antigens is impaired for several weeks. Induction of sensitization with the strong chemical stimulant 2,4-dinitrochlorobenzene is difficult. The lymphocyte proliferation response to stimulation with phytohemagglutinin is reduced, in the presence of a normal number of rosette-forming T cells (Fig. 5.2). This impairment is most marked soon after the appearance of the rash and improves progressively with passage of time thereafter. Coovadia *et al.* (1974) reported that the depression of immunity following measles was more severe than that in protein-calorie malnutrition. More recently, Coovadia *et al.* (1977) have shown that immunosuppression was more pronounced in children with measles who subsequently died than in those who recovered. The fatal outcome was associated with lower absolute counts of T and B lymphocytes, serum C3 concentration, and PHA-stimulated

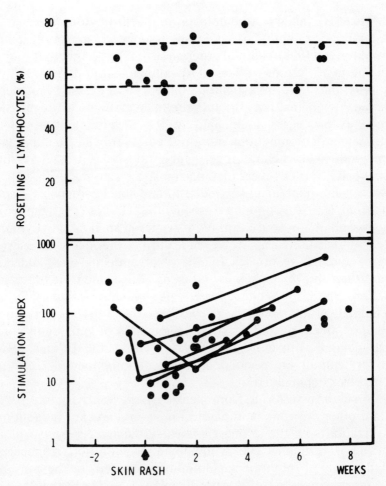

Figure 5.2. Depression of cell-mediated immunity associated with measles. The proliferative capacity of lymphocytes was reduced before, during, and for a short period after the appearance of measles rash. Stimulation index was calculated by cpm in lymphocyte cultures containing PHA/cpm in cultures without PHA. The number of T lymphocytes estimated by rosette formation with sheep red blood cells was generally normal or slightly low. The interrupted lines represent the proportion of T lymphocytes in the peripheral blood of healthy individuals (cpm = counts per minute).

lymphocyte proliferation. Similar transient alterations in immunocompetence may be observed after immunization with live attenuated measles virus vaccine. Kantor (1975) studied lymphocytes obtained at various intervals from individuals before and after immunization with vaccine containing measles, mumps, and rubella antigens, and stimulated with plant lectins phytohemagglutinin and pokeweed mitogen, and with unrelated antigen, *Candida albicans*. Mitogen-induced stimulation of DNA synthesis by lymphocytes was normal whereas in experiments with *Candida*, postvaccination lymphocyte samples showed a smaller increase in cell proliferation than specimens obtained before immunization. There was no alteration in the number of rosette-forming T lymphocytes. Cutaneous delayed hypersensitivity was not studied. However, Marks *et al.* (1974) showed that tuberculin reaction remains unaffected by immunization.

Infectious mononucleosis, a self-limited lymphoproliferative disorder caused by the Epstein–Barr virus, has also provided the interesting situation where delayed hypersensitivity and lymphocyte proliferative responses to plant mitogens, antigens, and allogeneic lymphocytes are depressed in the presence of an adequate number of morphologically normal T cells (Mangi *et al.*, 1974; Chandra, 1977i). In eight patients with infectious mononucleosis, Haider *et al.* (1973) observed negative tuberculin reactions during acute illness which became positive during convalescence.

Levene *et al.* (1969) described that a plasma factor in patients with active syphilis reduced lymphocyte transformation by phytohemagglutinin. Similar results were obtained by Thompson and Dwyer (1975). The inhibition was more pronounced when serum samples from patients with secondary syphilis were used, compared with primary syphilis. The inhibitory effect was not corrected by the addition of pooled normal serum. Interestingly, sera of patients with congenital syphilis enhanced lymphocyte proliferation, which was interpreted as being due to absence of a naturally occurring inhibitor or presence of stimulating factors.

5.3.1.1. Mechanisms of Anergy

The mechanisms of depression of cell-mediated immunity during and for a variable period after infection with viruses, bacteria, particularly intracellular invaders, fungi, and parasites, are not completely understood. Hypothetically, the organism might affect antigen-sensitive or effector lymphocytes, macrophages, or soluble mediators of the immune response at any of several steps in the sequence of events leading to delayed hypersensitivity reaction. Virus might directly damage or destroy T lymphocytes or a vigorous T cell response to a large concentration of the viral antigen may release soluble inhibitors of lymphocyte function that would then prevent response to other antigens. It is possible that some viruses preferentially stimulate a subpopulation of suppressor T lymphocytes. Another hypothesis postulates that the responder cells elaborate a chalone with a negative feedback function. These may be important modulating components of normal regulation in infective illness. The presence and persistence of intracellular pathogen or antigen enhances the occurrence and duration of anergy. Antigen elimination and clinical recovery are associated with termination of the anergic state. The anergy of miliary tuberculosis has been postulated to be due to extreme antigen excess which may bind all available receptors on lymphocytes, preventing subsequent interaction with intradermally administered antigen (Kent and Schwartz, 1967).

Besides cellular factors, serum samples of infected patients have an inhibitory effect on *in vitro* proliferative responses of lymphocytes of healthy donors (Chandra, 1977d, 1977f). The nature of the plasma inhibitor(s) is not clear. Suppressor factors have been demonstrated in the alpha and gamma globulins, albumin or prealbumin. Parts of organisms or substances released by them or by leukocytes and other body cells may be contributory. In some infections, antigen–antibody complexes may be inhibitory, possibly by nonspecific blockage of cell receptors. Many infections, including measles and chicken pox, are associated with elevated concentration of physiologically active free cortisol not bound with protein. Steroids are well-recognized immunosuppressive agents (Berenbaum, 1975). In addition, infection increases catabolism and nutritional requirements, and in some instances there are absolute losses of proteins and other nutrients. Dossetor and Whittle (1975) have shown the frequent occurrence of protein-losing gastroenteropathy in Nigerian children with measles. These changes in the nutritional milieu could contribute to impaired immune responses (Chandra, 1976b).

Experiments employing cell transfers in guinea pigs immunized with a variety of sensitizing antigens showed that anergy was due to an inhibitory environment since peritoneal exudate cells of anergic animals could transfer reactivity to nonimmunized recipients (Dwyer and Kantor, 1973). On the other hand, sensitization could not be transferred by injection of competent cells into anergic animals.

5.3.1.2. Clinical Significance

The clinical implications of infection-associated anergy are not known. Is there an increased incidence of infections and auto- and hetero-sensitization during and following immunosuppressive infections? Inactive tuberculosis may flare up after an episode of measles. Secondary infections are frequent complications after measles, influenza, and other viral and bacterial infections, but physical structural damage of mucous membranes and tissues would be as important here as immunosuppression. Primary sensitization to allergens can occur for the first time after an episode of acute infection. For example, hypersensitivity to cow's milk proteins may be initiated after a bout of gastroenteritis. The incidence and severity of atopic disease and IgE levels in patients with cystic fibrosis increase progressively with age (Chandra, 1977i). It is not known whether autoimmune disorders, immune complex disease, and other immunopathologic states may be initiated or precipitated by infection. In some situations, pathogens act as adjuvants providing exaggerated antigenic stimulus to elicit tissue-damaging antibodies.

5.3.2. Immunoglobulins and Antibodies

During infection, the synthesis of γ-globulins is increased severalfold (Cohen and Hansen, 1962). The number of circulating B lymphocytes is normal or increased (Coovadia *et al.*, 1974; Chandra, 1976b, 1977c). Serum concentrations of all major immunoglobulins are elevated (Table 5.2), principally as a result of enhanced synthesis of antibodies directed against the infectious

Table 5.2
Effect of Bacterial Infection and of Parasitic Infestation
on Serum Immunoglobulin Concentrations[a]

Group	IgG (mg%)	IgA (mg%)	IgM (mg%)	IgD (mg%)	IgE (U/ml)
Bacterial infection	2895 ± 256	220 ± 38	159 ± 33	11.1 ± 5.3	56 ± 37
Intestinal parasites[b]	1435 ± 216	180 ± 53	136 ± 47	7.6 ± 2.9	2410 ± 736
Healthy	1080 ± 192	110 ± 29	88 ± 21	1.9 ± 0.9	36 ± 21

[a] Values are given as mean ± standard deviation.
[b] *Ascaris* and/or hookworm.

agent. Nonspecific stimulus to formation of unrelated antibodies may also occur. Serum IgM level is increased in the acute phase as well as in chronic infections especially with intracellular pathogens, and IgA is predominantly elevated in chronic mucosal infections of the gastrointestinal and respiratory tracts. Increased IgE levels are characteristic of parasitic infestations (Johannsson *et al.*, 1968; Grove *et al.*, 1974; Chandra, 1977f) and vary with the severity of parasitic load and invasiveness of the pathogen.

The turnover of immunoglobulins is rapid and the plasma half-life of radiolabeled IgG is considerably shortened (Chandra, 1977c, 1977f). Antibody response to a challenge by an unrelated antigen is frequently depressed. For example, the titer of *Salmonella* agglutinins following primary and secondary immunization with typhoid-paratyphoid vaccine is significantly lower in infected individuals than in noninfected controls matched for nutritional status and several other parameters (Fig. 5.3).

5.3.3. Phagocytes

In infectious disease, most particularly of bacterial etiology, tissue macrophages and circulating microphages are essential protective mechanisms. Disease syndromes characterized by chronic infections have been related to neutrophil dysfunction. It is now clear that that the morphology and function of phagocytes may be significantly altered during episodes of infection. It is likely that alterations in macrophage and microphage function may influence other aspects of immune response, such as quantity and affinity of antibody produced, since phagocytes play a facilitatory role in many T cell- and B cell-mediated responses.

5.3.3.1. Number and Morphology

Infection is commonly associated with an increase in the proportion and absolute number of circulating polymorphonuclear leukocytes. In addition, a frequent occurrence is a "shift to the left," based on nuclear configuration, and defined by the presence of band forms of neutrophils, metamyelocytes, and sometimes myelocytes. The cytoplasmic changes include the presence of "toxic" granules which stain more prominently than granules of

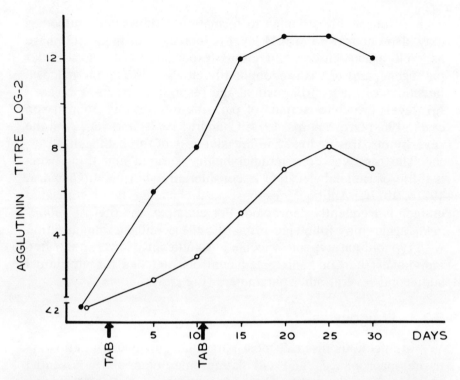

Figure 5.3. Effect of concomitant infection on *S. typhi* O agglutinin response. TAB = typhoid-paratyphoid A and B vaccine 0.1 ml intradermal injection; ○ = children with chronic pulmonary infection; ● = noninfected controls. From Chandra (1977f).

normal leukocytes, Döhle bodies consisting of light-blue amorphous inclusions, and cytoplasmic vacuoles (Ponder and Ponder, 1943). The severity of these morphologic alterations has been associated with a poor prognosis. Ultrastructurally, Döhle bodies are lamellar aggregates of rough-surfaced endoplasmic reticulum and toxic granules are peroxidase-containing lysosomes (McCall *et al.*, 1969).

5.3.3.2. Phagocytosis and Microbicidal Capacity

In patients with fulminant pyogenic infections, the ability of polymorphs to kill phagocytosed bacteria *in vitro* is impaired (Solberg and Hellum, 1972; Chandra, 1977f; Chandra *et al.*, 1977a). In these studies, the reduced granulocyte function was not the cause but the result of the infection, since leukocyte function returned to normal during recovery. Neutrophils with the characteristic toxic granules seen in large numbers in bacterial infections were found in one study to be slightly less efficient in dealing with ingested organisms than morphologically normal leukocytes (McCall et al., 1971). The reduction in total bacterial clearance by toxic neutrophils was associated with increases in both free and cell-associated bacteria suggesting a compromise in phagocytic as well as bactericidal activities. The infection-related abnormality of intracellular bacterial killing is dramatized in malnourished children in whom the presence of coexisting infection acts as a further depressive influence on bactericidal capacity of polymorphs (Chandra, 1977f; Chandra et al., 1977a).

Viruses may also impair neutrophil function. During the course of experimental infection with sandfly virus in seven volunteers, bactericidal activity was diminished by as much as 45% of baseline values, usually from the 7th to the 25th day of the illness (Bellanti et al., 1972). In a brief communication, Craft et al. (1976) reported diminished candidacidal activity of polymorphs obtained from children with respiratory syncytial virus or influenza virus infection. However, both these studies did not distinguish between inhibition of ingestion and failure of intracellular killing, nor did they look at the infection-induced nutritional and metabolic changes which may suppress leukocyte function. A marked reduction in leukocyte glucose-6-phosphate dehydrogenase and in quantitative nitroblue tetrazolium dye reduction suggested that intracellular microbicidal pathways may be impaired (Bellanti et al., 1972). The ability of polymorphs to ingest staphylococci was inhibited when they were incubated with influenza virus. It was postulated that the inhibition of phagocytosis was the result of the viral action at the cell surface (Larsen and Blades, 1976).

5.3.3.3. Mobilization and Chemotaxis

The marginal pool of neutrophils is unchanged during infection but the marrow reserves are significantly depleted (Chandra et al., 1976a). Circulating microbial products including endotoxin in such individuals provide a continuous stimulus for the release of granulocytes from the bone marrow.

Directed movement of polymorphs *in vitro* toward a chemotactic source is reduced during infectious illness (Rosen *et al.*, 1975; Chandra *et al.*, 1976a). Toxic neutrophils from patients with severe sepsis had 21% of the chemotactic activity of controls (McCall *et al.*, 1971). Anderson *et al.* (1976) found both random mobility and chemotactic ability of neutrophils from patients with uncomplicated measles to be grossly impaired. This was confirmed *in vivo* by abnormal Rebuck skin windows. The abnormalities of motility reversed completely to normal by the 11th day after the onset of the rash. Endotoxin-activated generation of chemotactic activity was normal in the serum, which did not contain leukotactic inhibitors either.

5.3.3.4. Biochemical Constituents and Metabolic Response

Bacterial infection is associated with a marked decrease in intraneutrophilic concentration of lysozyme (Hansen and Anderson, 1973; Chandra *et al.*, 1977b) which spills into the plasma pool. The ratio of plasma lysozyme to neutrophil lysozyme is increased 5-fold during infection (Chandra *et al.*, 1977b). In a sequential study of intraneutrophilic levels of lysozyme, myeloperoxidase and lactoferrin in acute bacterial infection, Hansen *et al.* (1976) found decreased values during the first week of illness followed by a slow increase over the next two weeks. Nadir values coincided with maximal toxic granulation of the granulocytes. It was postulated that infection results in deficient synthesis of leukocyte antibacterial proteins in the bone marrow. In a sequential study of neutrophil biochemical functions in volunteers infected with sandfly fever virus, Bellanti *et al.* (1972) found a marked reduction in glucose-6-phosphate dehydrogenase activity and impaired nitroblue tetrazolium dye reduction, both of which returned to normal within 3–4 weeks. No changes were detected in leukocyte 6-phosphogluconic dehydrogenase activity.

6

IMMUNOCOMPETENCE IN UNDERNUTRITION

There is an impressive mass of published evidence which suggests that nutritional status conditions the host to infectious disease. Nutritional deficiencies generally decrease the capacity of man to resist the occurrence and consequences of infection. This dynamic interaction between nutrition and infection will be altered by the type, inoculum size, and metabolic requirements of the pathogenic organism, and by the nature and severity of dietary deficiencies. In clinical practice, one usually encounters multiple nutrient deficits so that in man the effect of the lack of specific dietary elements on various facets of the immune response is difficult if not impossible to assess. The precise definition of both nutritional and immunologic profile in individuals examined is important so that the interactions of one upon the other is critically assessed and valid conclusions are drawn.

6.1. LYMPHOID TISSUES

Gross and histomorphologic changes in lymphoid organs of severely malnourished children and adults have been documented for almost half a century (Vint, 1937). The central organs of the immune system, notably the thymus, as well as the peripheral tissues, for example the spleen and lymph nodes, are significantly altered in size, weight, architecture, cellular components, and fine structure.

6.1.1. Thymus

Thymuses of undernourished children are consistently small and on an average weigh less than one-half of the normal organ (Watts, 1969; Smythe *et al.*, 1971). Occasionally, the reduction in thymus weight is extreme, the organ weighing only a few milligrams, which has led to the term *nutritional thymectomy,* a phenomenon dramatically illustrated in experimental deprivation of calories and essential nutrients in laboratory animals.

Smythe *et al.* (1971) conducted postmortem studies on 23 children dying of marasmus and 47 children with kwashiorkor. The mean thymic weight, expressed as a percentage of organ weight for body length, was 37% in marasmus and 30% in kwashiorkor. Two types of histologic abnormalities were distinguished. One-fourth of the specimens showed acute involution characterized by lobular atrophy with loss of corticomedullary differentiation and reduction in the number of thymocytes. In the majority, histologic evidence of chronic atrophy consisting of narrow, withered lobules with indented edges, absence of corticomedullary differentiation, scanty thymocytes, and prominent fibroblast-like cells, was seen. The mean thickness of thymic lobule was 476 μm in marasmus, 538 μm in kwashiorkor, and 1172 μm in the well-nourished group. In the marasmic group, 8% of the organs had normal microscopic structure.

In the autopsy data reviewed by Purtilo and Connor (1975), the thymus gland of malnourished children ranged in weight from 2.5 to 28 g, and averaged 5 g. There was loss of corticomedullary demarcation, lymphocytic depletion, and relative increase in fibrovascular tissue (Fig. 6.1). The Hassall bodies may be normal, dilated, degenerated, or even calcified (Fig. 6.2). The crowding of Hassall corpuscles so characteristic of secondary involution of the thymus is in marked contrast to their absence in primary thymic dysplasia. The histomorphologic changes in the thymus are the severest in patients with antemortem lymphopenia. The simultaneous occurrence of overwhelming infection, most particularly measles, aggravates the thymic involution. However, similar histologic lesions and impaired cellular immunity are seen in noninfected children with energy–protein undernutrition, which suggests that thymolymphatic atrophy and hypofunction is the

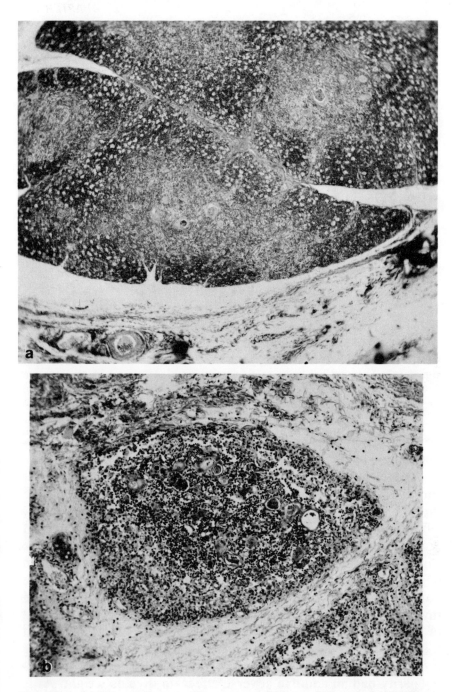

Figure 6.1. (a) Normal thymus. The cortex and the medulla are clearly distinguished. ×40. (b) Thymus in energy–protein undernutrition. There is loss of cortico-medullary demarcation, depletion of lymphocytes and increase in fibrovascular tissue. ×60. Courtesy of Dr. David T. Purtilo.

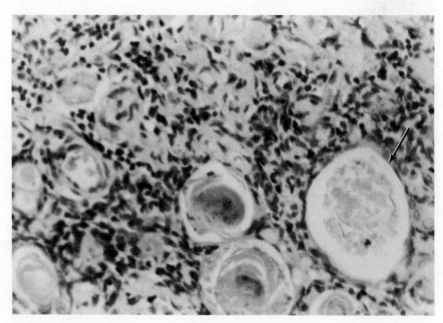

Figure 6.2. Thymus in energy–protein undernutrition. Hassall corpuscles are crowded together and some are in the process of dilatation and degeneration (arrow). ×250. Reduced 30% for reproduction.

primary lesion in nutritional deficiency and precedes the compounding complication of infection.

Adverse nutritional factors during intrauterine development can affect the growth of fetal lymphoid organs, particularly the thymus. In autopsy studies conducted by Naeye *et al.* (1971, 1973) undernutrition was identified as a cause of low birth weight in poor urban mothers. Body length and all organ weights were smaller and thickness of abdominal subcutaneous fat less in infants from families whose income was close to or below the poverty line, irrespective of racial background. Weights of the thymus and spleen were disproportionately smaller than weights of other organs. A progressive decrease in newborn organ weight took place from overweight mothers with high weight gain during pregnancy to underweight mothers with low weight gain. Increasing parity had an additive adverse effect on thymus weight. As is probably true of general body growth and brain size, reduction in the size and function of the fetally involuted thymus may have long-lasting effects (Chapter 8).

6.1.2. Lymph Nodes

The nutritional lesion in the lymph node is most pronounced in the paracortical thymus-dependent regions. Lymphocytes are sparse and germinal centers are small and decreased in number (Fig. 6.3). Plasma cells and macrophages are relatively increased.

The mesenteric lymph nodes may not be enlarged despite recurrent and chronic gastrointestinal infections, with the exception of tuberculosis which often presents as abdominal lymphadenopathy.

6.1.3. Spleen

The size and weight of the spleen in human malnutrition varies from large to normal to small. In one study, the average spleen weight was 54% and 70% of the normal in kwashiorkor and marasmus respectively (Smythe et al., 1971). Histologically, the reduction in germinal center activity is striking. There is marked depletion of lymphocytes in the thymus-dependent periarteriolar regions (Fig. 6.4). There is a relative distension of sinuses which are filled with red blood cells. The number of cells in mitosis is reduced, which parallels the observation of reduced incorporation of tritiated thymidine by lymphoid tissues of nutritionally deprived laboratory animals. These changes in lymphoid populations in the spleen are more pronounced in young children dying of severe infections, particularly measles.

6.1.4. Gut-Associated Lymphoid Aggregates

There is very poor documentation of histomorphologic alterations in the gut-associated lymphoid tissues in malnutrition. The tonsils and adenoids are often small to the point of being vestigial. This is evident clinically (Smythe et al., 1971; Chandra, 1972) as well as at autopsy. There is marked depletion of lymphocytes, particularly in the paracortical areas, and reduction in the number and size of germinal centers (Fig. 6.5). The average tonsillar area measured by multiplying length and depth was 23 mm^2, 22 mm^2, and 44 mm^2 in marasmus, kwashiorkor, and controls respectively (Smythe et al., 1971).

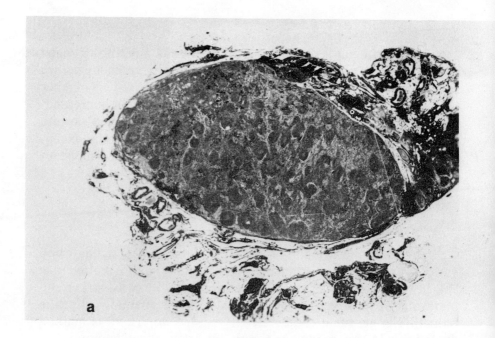

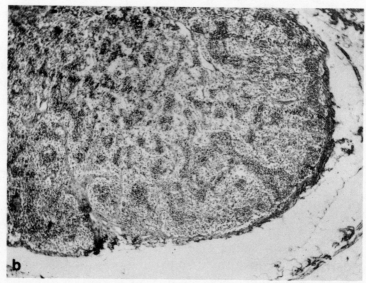

Figure 6.3. (a) Normal lymph node. The germinal centers are prominent. ×20. (b) Lymph node in energy-protein undernutrition. The germinal centers are scanty and small. There is depletion of the paracortical lymphocytes. ×40. Courtesy of Dr. David T. Purtilo.

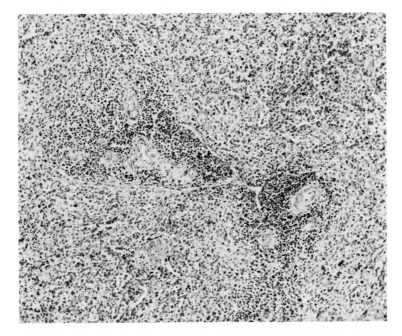

Figure 6.4. Spleen in energy–protein undernutrition. There is marked atrophy of periarteriolar lymphocytic sheath. ×60.

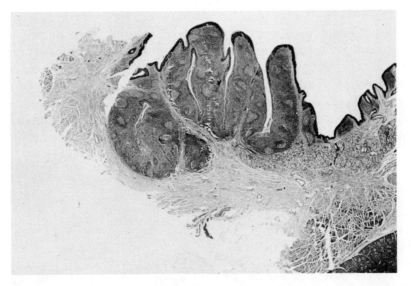

Figure 6.5. Tonsil in energy–protein undernutrition. There is marked involution of lymphoid component. ×20. Courtesy of Dr. David T. Purtilo.

Lymphoid aggregates in the small intestine, appendix, and colon are reduced in size and lymphocyte population. Smythe *et al.* (1971) measured the depth of the lymphoid tissue in small intestine and found that the mean thickness of Peyer's patch was 405 μm, 575 μm, and 765 μm in kwashiorkor, marasmus, and well-nourished groups, respectively. The mean size of lymphoid tissue in the appendix was 360 μm, 405 μm, and 675 μm, respectively.

6.1.5. Pathogenesis

The pathogenesis of these morphologic alterations in the thymus and other organs of the immune apparatus may lie in one or more of several mechanisms. In energy-protein undernutrition, cell division and proliferation are severely restricted consequent upon reduced availability of nutrients. Nutritional deficiency is associated with changes in the production and concentration of several hormones. Many hormones, including corticosteroids, adrenaline, insulin, and thyroxine may affect immunocompetent leukocytes (Claman, 1972; Farid *et al.*, 1976; Eriksson and Hedfors, 1977). Plasma cortisol level may be elevated (Rao *et al.*, 1968) and in kwashiorkor, the reduction in serum albumin frees up a larger proportion of cortisol for action on tissues (Chandra, 1977h). "Stress" may result in additional changes in hormone levels. It has been suggested that tissues with a high nucleus to cytoplasm ratio, such as lymphoid organs, show a preferential uptake of corticosteroids with resultant cell lysis, involution, and hypofunction. The lympholytic and immunosuppressive action of corticosteroids is established (Berenbaum, 1975). Endotoxemia is frequently associated with sepsis in malnutrition and may contribute to lymphoid atrophy. Finally, some pathogenic organisms may invade lymphoid tissues and cause direct cellular damage.

6.2. CELL-MEDIATED IMMUNITY

6.2.1. Leukocyte Counts

There is a wide scatter of values reported for total leukocyte number in the peripheral blood of children with nutritional deficiencies, from low to normal to high count. Most often, the total white cell number is mildly increased. In observations made on man, the strict differentiation between the effects of nutritional deficiency and of frequently associated infection is not possible. Sepsis usually produces leukocytosis but fulminant bacteremia can depress the bone marrow and lead to leukopenia.

The proportion and absolute number of lymphocytes is generally normal. About 10–30% of undernourished individuals show mild to moderate reduction in lymphocyte counts. Lymphopenia, defined as an absolute lymphocyte count less than 2500/mm^3, was seen in 15% of young Indian patients with energy-protein undernutrition (Chandra, 1972). Superimposed infection, particularly if it is acute and fulminant, can further depress the number of circulating lymphocytes. In two studies from South Africa, lymphopenia was observed in one-sixth and two-thirds of infected infants with kwashiorkor (Smythe et al., 1971; Rosen et al., 1975). Neumann et al. (1975) found profound lymphopenia (cell count below 1000/mm^3) in 9% of children with severe malnutrition.

6.2.2. Lymphocyte Subpopulations

The consistent observation of impaired delayed hypersensitivity responses in undernutrition and the development of the techniques of the morphologic identification of various subsets of lymphocytes led to the logical study of the proportion and absolute number of lymphocyte subpopulations in the peripheral blood and its correlation with nutritional status and with *in vitro* and *in vivo* indices of cellular immunity. Initial studies from India and Ghana showed that the frequency of T cells in the peripheral blood was much reduced in malnutrition (Chandra, 1974a; Ferguson et al., 1974). This reduction in T lymphocytes paralleled the severity of weight loss, impaired cutaneous delayed hypersensitivity response to 2,4-dinitrochlorobenzene, and decreased DNA synthesis by lymphocytes stimulated with phytohemagglutinin (PHA) (Chandra, 1974a). The abnormalities were quickly and completely reversed on nutritional improvement, thereby ruling out any primary defect of the thymus. The dramatic increase in T cell number following the provision of nutritional supplements occurs within a few days (Fig. 6.6) and may well be seen before any measurable improvement in clinical features (weight, mid-arm circumference, skin-fold thickness) or biochemical indices (serum proteins and albumin concentrations) of nutritional status. The reduction in T lymphocytes has been confirmed in studies from Nepal and Calcutta (Bang *et al.*, 1975), the Ivory Coast (Schopfer and Douglas, 1976a), Thailand (Kulapongs *et al.*, 1977a) and Guatemala (Keusch *et al.*, 1977a).

Similar changes in the proportion and absolute number of T lymphocytes are seen in small-for-gestation, low-birth-weight infants, a syndrome of fetal growth retardation of diverse etiology. Complementary data from Los Angeles and New Delhi attest to

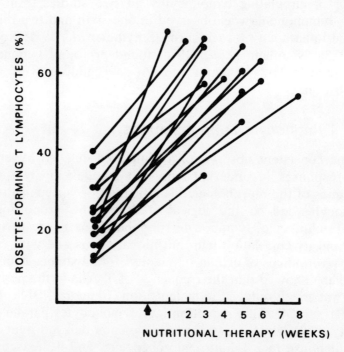

Figure 6.6. Proportion of T lymphocytes in the peripheral blood of patients with energy-protein undernutrition. The speed of recovery of T cell number after nutritional supplementation is shown. This index of cellular immunity was seen to return to the normal range of values in 1-4 weeks, often before any significant change in clinical or biochemical parameters of nutrition. Based on the data of Chandra, 1974a.

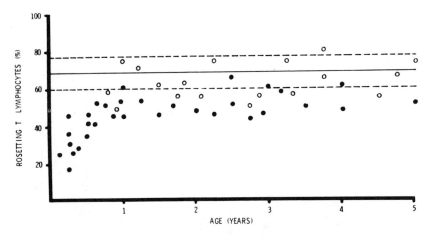

Figure 6. 7. Rosette-forming T lymphocytes in small-for-gestation infants examined several months after birth. In infants and children with persistent growth retardation, ●, the proportion of T cells was significantly lower and correlated with impaired delayed hypersensitivity. The range of normal values is shown as horizontal lines. From Chandra et al., 1977c.

the reduction of T cells in the peripheral blood of infants with intrauterine growth retardation (Ferguson et al., 1974; Chandra, 1975c). A follow-up study in Newfoundland revealed that the percentage of rosette-forming T lymphocytes in the blood remains lower than expected for several months after birth (Fig. 6.7) (Chandra et al., 1977c). This correlated with impaired cutaneous delayed hypersensitivity responses to a battery of ubiquitous antigens (Chandra, 1977f). In children with persistent retardation of growth measured by weight for height age, the depression of cell-mediated immunity was observed for as long as 5 years after birth (Chandra et al., 1977c).

The influence of various dietary nutrients on the thymus and T lymphocytes is not clear. Iron is believed to have a direct effect on the integrity of lymphoid tissues, a thesis supported by experimental data (Chandra et al., 1977d). Preliminary studies reported a mild to moderate reduction in the number of rosette-forming lymphocytes in iron deficiency anemia (Chandra, 1975d, 1976d; Bhaskaram and Reddy, 1975; Chandra et al., 1977d). It is necessary to obtain similar information in other deficiency syndromes.

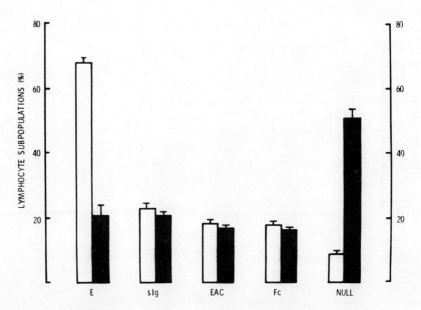

Figure 6.8. Lymphocyte subpopulations in children with energy–protein undernutrition (solid bars) and well-nourished controls (open bars). There was a marked reduction in rosetting T lymphocytes and a proportionate increase in null cells. The percentage of null cells was derived by substracting from 100 the proportion of E-rosetting and sIg-bearing lymphocytes. E = E-rosetting T cells, EAC = antibody-complement-rosetting cells, sIg = surface-immunoglobulin-bearing B cells, Fc = Fc-receptor-bearing lymphocytes. From the data of Chandra, 1977a.

It is not established if the reduction in circulating T lymphocytes in energy–protein undernutrition reflects an absolute decrease in the total body pool of these leukocytes or merely a sequestration and diversion into other areas. The mechanisms leading to thymic atrophy have been discussed earlier.

The percentage of B lymphocytes in the peripheral blood in malnutrition is normal or increased (Chandra, 1977a, 1977f). The increase is often seen in those children who have an obvious infection associated with nutritional deficiency (Bang *et al.*, 1975).

If T lymphocytes are reduced in proportion and B lymphocytes are normal or only marginally increased, what are the morphologic and functional attributes of the remaining cells? The lymphocytes which do not bear the conventional markers for T or B cells have been given the rather unsatisfactory label of *null* cells. The relative proportion of lymphocytes with various cell surface markers in undernourished and well-nourished children is shown in Figure 6.8. The major differences are in the numbers of T lymphocytes and null cells. A small proportion of null cells showed surface receptors for the Fc part of IgG and for C3.

The functional significance of the null cells is not completely understood. However, preliminary data suggest that they have a suppressor influence on mitogen-stimulated DNA synthesis in normal T lymphocytes (Fig. 6.9). Additionally, such null cells

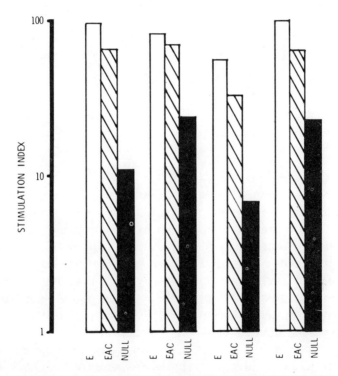

Figure 6.9. DNA synthesis by T lymphocytes of a healthy donor in the presence of mitomycin-treated lymphocyte subpopulations of four malnourished children. T lymphocytes were obtained by E rosetting, B lymphocytes by EAC rosetting, and null cells by passage of mononuclear layer through anti-Ig coated nylon wool column, E rosetting of eluted cells, density gradient centrifugation, and collection of interphase layer. DNA synthesis in response to PHA was measured by 6-hour ^{3}H-thymidine uptake. Stimulation index was calculated as ratio of cpm PHA-containing culture/cpm culture without PHA. From Chandra, 1977a.

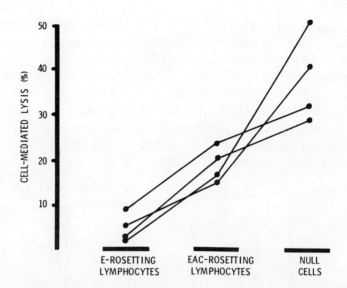

Figure 6.10. Xenogeneic target cell lysis by lymphocyte subpopulations of four malnourished children. Cell populations were separated as in Figure 6.9 and layered over labeled DBA/2 fibroblast culture. Cell damage was measured by isotope release at the end of 24-hour incubation. From Chandra, 1977a.

exert a "spontaneous" (without the obligatory presence or participation of antibody) cytotoxicity against xenogeneic target cells, a capacity much greater than shown by B lymphocyte-enriched or T lymphocyte-enriched cell preparations (Fig. 6.10). The biological importance of these *in vitro* observations is moot. The possibility of such reactions subserving a negative feedback in many immune responses, physiological and pathological, must be considered. Patients suffering from several disorders with the common denominator of depressed cell-mediated immunity, including lepromatous leprosy, primary immunodeficiency, and lymphoreticular malignancy, have an increased number of null cells.

6.2.3. Delayed Hypersensitivity

Cell-mediated immunity is often assessed by skin test reactions to a battery of antigens which the recipients are likely to have met earlier in life; for example, streptococcal antigens (streptokinase and streptodornase), *Candida albicans*, purified protein derivative (PPD) of tuberculin, trichophyton, and mumps. Alternatively, sensitization may be deliberately attempted by the use of keyhole limpet hemocyanin, or 2,4-dinitrochlorobenzene. The antigens are injected intradermally and induration observed 48–72 hours later is considered a positive response. Skin tests have the merits of being *in vivo* reactions, ease of administration and interpretation, comparability, and repeatability, important considerations in field studies and in situations where laboratory facilities are minimal. However, subtle differences are not as reliable as in adults. Also, macrophage function as well as individual variation in nonimmunologic skin reactivity, as assessed by response to chemical irritants, contributes to the final skin reaction. Structural, hormonal, and biochemical changes in the skin in nutritional deficiency can influence the hypersensitivity response.

The delayed hypersensitivity response is a composite of three distinct sequential processes. The afferent limb involves sensitization of T lymphocytes against a macrophage-processed antigen. The efferent limb is characterized by the production of soluble chemical mediators or lymphokines and comes into play when sensitized T cells recognize and interact with the intradermally injected antigen. The inflammatory response, probably triggered by lymphokines released at the local skin site and chemotactic factors, produces the dermal induration characteristic of a positive reaction. Nutritional deficiency depresses one or more components of the delayed hypersensitivity response.

Early observations by Jayalakshmi and Gopalan (1958) and Harland (1965) showed that children with subnormal rate of growth due to dietary protein deficiency had an impairment of their delayed hypersensitivity response to tuberculin. The reduction in response was directly related to the severity of weight deficit and was not absolute since 50 T.U. or higher dose of tuberculin overcame the anergy. A short period of good nutrition through high-protein (4g/kg) diet repaired the delayed hypersensitivity response.

Recent studies employing a battery of antigens have confirmed the frequent occurrence of anergy in energy–protein undernutrition (Chandra, 1972, 1974b, 1977d; Neumann *et al.*, 1975).

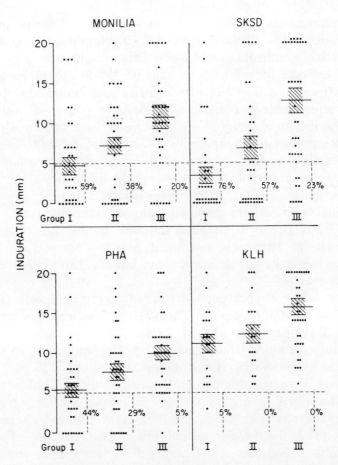

Figure 6.11. Cutaneous delayed hypersensitivity to monilia, streptokinase-streptodornase (SKSD), phytohemagglutinin (PHA), and keyhole limpet hemocyanin (KLH), related to nutritional status. Group I—severe malnutrition, weight 51%–60% of the 50th percentile Harvard standard and/or serum albumin <2.5 g%, and clinical features characteristic of kwashiorkor or marasmus. Group II—moderate malnutrition, weight 61%–80% of the standard. Group III—apparently healthy controls with weight ≥ 81% of the standard. From Neumann et al., 1975.

Cutaneous delayed hypersensitivity was reduced both in terms of decreased diameter of induration and percentage of nonreactors. Anergy was more pronounced in children with severe malnutrition (Fig. 6.11). The failure to produce immunization with the strong sensitizing agents 2,4-dinitrochlorobenzene and 2,4-dinitrofluorobenzene in malnourished children (Smythe *et al.*, 1971; Chandra, 1972, 1974a; Feldman and Gianantonio, 1972; Edelman *et al.*, 1973; Schlesinger and Stekel, 1974; Smith *et al.*, 1977) and malnourished adults (Law *et al.*, 1973) reflects the magnitude of depression of cell-mediated immunity in nutritional deficiency. If malnutrition occurs early in the first year of life, cutaneous anergy may persist for a long time (Dutz *et al.*, 1976).

A failure of response to Mantoux tuberculin test in some patients with active disease with *Mycobacterium tuberculosis* is known. It appears likely that nutritional status is a critical variable in depressed delayed hypersensitivity. In a group of patients diagnosed to have active pulmonary tuberculosis on the basis of radiologic findings and sputum bacteriology, tuberculin reactivity correlated significantly with serum transferrin concentration (Fig. 6.12). Transferrin levels are a sensitive biochemical

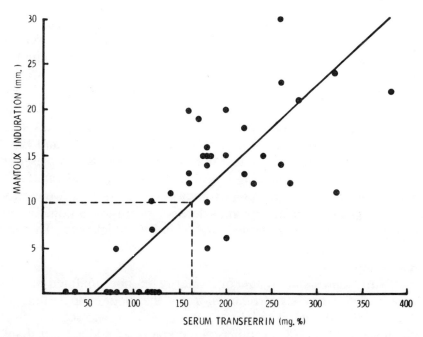

Figure 6.12. Cutaneous delayed hypersensitivity to 5 T.U. tuberculin related to serum transferrin concentration in patients with pulmonary tuberculosis. A positive Mantoux test, induration > 10 mm, was generally observed in individuals with serum transferrin level of ≥162 mg%. Patients studied through the courtesy of Dr. H. B. Dingley, Tuberculosis Hospital, New Delhi.

index of energy–protein undernutrition (Chapter 3). Similar data have been reported by Harrison *et al.* (1975).

Two-thirds of Nigerian adult patients with sputum-positive pulmonary tuberculosis had a negative skin reaction to 4–5 T.U. of PPD. The size of the reaction correlated significantly with serum concentrations of albumin and transferrin. The cell-mediated immune response was generally depressed. An important observation in this study was the slower response to treatment shown by the more severely undernourished patients, supporting the role of malnutrition in infection immunity.

Tuberculin reactions sometime after the administration of B.C.G. vaccine can be used as an index of cellular immunity. Tuberculin conversion after immunization was significantly less frequent in children with postnatally acquired malnutrition (Chandra, 1972; Abbassy *et al.*, 1974a) as well as in a group of infants with intrauterine growth retardation (Manerikar *et al.*, 1976). In a study on Nepali village children, the average induration of PPD–S and PPD–B tuberculin reaction was considerably bigger in well-nourished children compared with the undernourished group (Ziegler and Ziegler, 1975). An analysis of the data by the number and proportion of children who had post-B.C.G. induration greater than 4 mm showed similar differences in the two groups. An improvement in nutritional status was associated with increased reactivity. The impairment of B.C.G. immunity was seen in all grades of severity of nutritional deficiency. Sinha and Bang (1976) found the tuberculin test response to be grossly impaired in kwashiorkor and marasmic kwashiorkor, and adequate in marasmus. The nutritional categorization was based on weight for age and calculation of arm muscle and fat cross-sectional areas based on arm circumference and skin-fold thickness. Therapy with massive doses of vitamin A did not influence post-B.C.G. tuberculin sensitivity.

Nutrition-related cutaneous anergy is not confined to the severe deficiency syndromes nor to the poor populations of developing countries. Kielman *et al.* (1976) showed a lower incidence of post-B.C.G. tuberculin conversion and smaller diameter of induration on Mantoux test in a group of rural infants and children with moderate energy–protein undernutrition. Bistrian *et al.* (1975) have reported a high frequency of adult protein–calorie malnutrition and visceral attrition as reflected by decreased levels of serum albumin and transferrin among hospitalized patients in urban Massachusetts. Lymphocyte count was reduced and cell-mediated immunity impaired as measured by the Catalona technique for contact sensitization with 2,4-dinitrochlorobenzene and delayed hypersensitivity to *Candida*. The mechanisms of production of nutritional deficiency in hospitalized patients are not clear. The malnourished state was thought to be a consequence of low intake and of catabolic response to stress. The administration of intravenous glucose, a frequent hospital practice, may induce insulin release and reduce the release of amino acids from muscle, and thereby their availability for visceral protein synthesis.

Cutaneous delayed hypersensitivity is impaired in iron deficiency with or without anemia (Chandra, 1975d, 1976a; Mac-Dougall *et al.*, 1975; Chandra *et al.*, 1977d). Folate deficiency can be associated with failure to mount hypersensitivity reaction to 2,4-dinitrochlorobenzene (Gross *et al.*, 1975). It is likely that several other nutrients singly or in combination influence cell-mediated immunity, as observed in experimental animals.

6.2.4. Lymphocyte Proliferation

The capacity of lymphocytes to synthesize DNA in response to mitogenic or antigenic stimulus is an *in vitro* correlate of cell-mediated immunity. Several studies have documented a reduction in lymphocyte reactivity to the T cell mitogen PHA in chronic undernutrition (Geeshuysen *et al.*, 1971; Smythe *et al.*, 1971; Chandra, 1972, 1974a, 1978a; Grace *et al.*, 1972; Sellmeyer *et al.*, 1972; Neumann *et al.*, 1975; Schopfer and Douglas, 1976a; Kulapongs *et al.*, 1977a) and in acute starvation (Holm and Palmblad, 1976). Even though there is a wide scatter of values obtained for lymphocyte stimulation and considerable overlapping of individual responses in groups based on nutritional status, the mean stimulation index in undernourished subjects is significantly lower than the corresponding index in the well-nourished (Neumann *et al.*, 1975). Mitosis is similarly depressed. Rarely, lymphocyte transformation response to PHA may be normal (Moore *et al.*, 1974; Schlesinger and Stekel, 1974). The unstimulated lymphocytes of malnourished children have a higher DNA content (Smythe *et al.*, 1971; Chandra, 1978a), a consequence of stimulation *in vivo* due to concurrent infection. Lymphocyte stimulation *in vitro* in response to specific antigens PPD, streptolysin O, and tetanus toxoid was diminished (Jose *et al.*, 1970; Chandra, 1978a). On the other hand, the responses to influenza A2 antigen and streptococcal Group A polysaccharide were enhanced. Lymphocyte proliferation in the presence of pokeweed mitogen is normal or enhanced (Schopfer and Douglas, 1976a).

If cell-mediated immunity *in vivo* and *in vitro* is consistently depressed in malnutrition, one would like to define the critical dietary elements that are responsible for it, since energy–protein undernutrition is the sum total of deficiencies of several nutrients. In man, there are obvious difficulties in attributing defects in immunity function to the lack of specific nutrients. There is some data on iron- and folate-deficient individuals. The *in vitro* transformation response to mitogens and antigens is reduced in iron-deficient individuals (Chandra, 1975d, 1976a; MacDougall *et al.*, 1975; Bhaskaram and Reddy, 1975). The critical etiopathogenetic factor is the lack of iron and not anemia, since iron-deficient subjects with normal concentration of hemoglobin may also show lymphocyte dysfunction. There is a significant correlation between depressed *in vitro* response and transferrin unsaturation (Fig. 6.13) (Chandra *et al.*, 1977d). Reduced DNA synthesis in the bone marrow (Hershko *et al.*, 1970) and in lymphocyte cultures (Joynson *et al.* 1972) has been documented in iron deficiency. Impaired cellular immunity *in vitro* improved dramatically after administration of iron (Chandra, 1975d, 1977f). On the other hand, Kulapongs *et al.* (1974) examined 8 Thai infants with severe iron-deficiency anemia and did not observe any evidence of immunosuppression.

Deficiency of folates is also associated with impaired cellular immunity (Gross *et al.*, 1975). The lack of vitamin B_{12} may or may not be associated with impaired cell-mediated immunity. Mac-Cuish *et al.* (1974) observed that patients with pernicious anemia showed significant depression of lymphocyte transformation to three different doses of PHA. Radioautographic examination of PHA-stimulated cells indicated that the results were due to a failure of intranuclear incorporation of ^{3}H-thymidine by the patients' lymphocytes rather than to a failure of PHA to induce

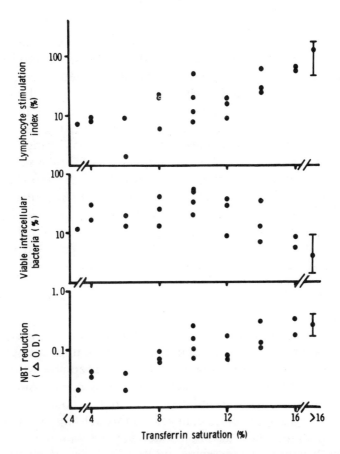

Figure 6.13. Lymphocyte proliferation, intracellular bactericidal capacity of neutrophils, and quantitative nitroblue tetrazolium test, related to serum transferrin saturation. The bars indicate the means and ranges of values obtained in iron-replete controls. From Chandra et al., 1977d.

blastogenesis. The number of T cells was only very slightly lower in pernicious anemia patients.

6.2.5. Methodological and Pathogenetic Considerations

The techniques employed for evaluation of cell-mediated immunity are critical variables of results obtained. Valid comparisons between the data of two studies can be made only if identical test procedures were used. The limitations of the cutaneous delayed hypersensitivity tests have been discussed earlier. The method of obtaining and separating leukocytes, duration of lymphocyte culture, dose of mitogen or antigen, method of assessing proliferative capacity, are all important in *in vitro* lymphocyte stimulation studies. The numerous modulating factors involved in human T lymphocyte–sheep red blood cell rosetting phenomenon may explain the discrepencies between "normal" range obtained in different laboratories. Are all lymphocytes forming nonimmune contact with one or more sheep erythrocytes thymus-dependent cells? In studies by Schopfer and Douglas (1976a) in the Ivory Coast, the absolute number of T lymphocytes was lower in kwashiorkor children. Lymphocytes forming "large" (3 or more than 3 sheep red blood cells) rosettes were fewer. On the other hand, if all cells forming nonimmune contact with one or more sheep erythrocytes were included, the percent of rosetting lymphocytes was comparable in well-nourished and kwashiorkor children.

The depression in lymphocyte number and proliferation in malnutrition could result from one of several pathogenetic factors: reduction in number of T lymphocytes; impaired ability to synthesize DNA; increase in "null" cells possessing suppressor activity; cytophilic inhibitors derived from microbes, host tissues, IgE, hormones, chalones, etc.; changes in the activity of membrane-bound enzymes such as cholinesterase; activation of latent viruses with immunosuppressive ability. The sera of malnourished children frequently inhibit lymphocyte DNA synthesis (Moore *et al.*, 1974). Infection often associated with nutritional deficiency contributes to the development of serum inhibitors including microbial products, cortisone, chalones, acute-phase reactants, IgE, and other unknown factors. Mortensen *et al.* (1975) have reported that purified human C-reactive protein binds selectively to human T lymphocytes, inhibits their ability to form spontaneous rosettes with sheep red cells, and depresses lymphocyte response to allogeneic cells in mixed leukocyte cultures. However, C-reactive protein did not impair PHA-induced blastogenesis. α-Fetoprotein is elevated in small-for-gestation low-birth-weight infants and may contribute to impaired cell-mediated immune responses (Chandra *et al.*, 1976c; Chandra and Bhujwala, 1977).

6.3. IMMUNOGLOBULINS AND ANTIBODIES

6.3.1. γ-Globulin Concentration and Metabolism

Low serum level of albumin is a characteristic feature of kwashiorkor and the nutritional edema syndrome. Hypoalbuminemia provides an underestimate of the extent of depletion of the body albumin, since reduction of extravascular pool is proportionally greater than that of the circulating albumin. Predominantly in calorie deficiency of which marasmus is the extreme example, albumin concentration is frequently normal or only slightly reduced. Irrespective of total proteins or other fractions, the levels of γ-globulins are normal or more often increased, the latter attributable to frequent associated infections. These alterations in serum protein components have been estimated by protein fractionation using biochemical and electrophoretic techniques (Anderson and Altman, 1951).

Cohen and Hansen (1962) used two isotopes with different emission spectra to measure simultaneously the turnover of serum albumin and γ-globulin. The total body pool of albumin and its synthesis rates were reduced. The fractional breakdown rate was low. In uninfected children with kwashiorkor the distribution and turnover of γ-globulin was relatively unaffected, indicating that plasma cells were able to utilize available amino acids preferentially. In cases of kwashiorkor complicated by obvious infection, the γ-globulin synthesis rate was higher (mean 163 mg/kg/day) and the catabolic rate was slightly lower. The role of environmental factors, particularly infection, in affecting these alterations was shown by a 3-fold increase in γ-globulin synthesis rates in West African adults living in hyperendemic Gambia (mean 169 mg/kg/day) compared with the same ethnic group in the United Kingdom (mean 50 mg/kg/day). Nutritional recovery was associated with 3-fold increase in synthesis rate and considerable repletion of the total body albumin, which was proportionally greater in the extravascular pool. In contrast, the average γ-globulin level and synthesis rate did not alter during refeeding, except in malnourished subjects with infections.

Kwashiorkor is often complicated by protein-losing gastroenteropathy (Cohen *et al.*, 1962), which may be the result of morphological alterations in the intestinal mucosa (Amin *et al.*, 1969) or of associated infections, such as measles (Dossetor and Whittle, 1975). Intravenously administered ^{131}I-γ-globulin was found in fecal specimens and was protein bound since the label in saline extracts of stools was almost completely precipitated by 10% trichloroacetic acid (Cohen and Hansen, 1962).

6.3.2. Immunoglobulin Levels and Turnover

Serum immunoglobulin concentrations reflect the past infectious experience of the individual. In the majority of children with energy–protein undernutrition, serum levels of IgG, IgA, and IgM are normal or elevated (Table 6.1) (Najjar *et al.*, 1969; El Gholmy *et al.*, 1970; Alvarado and Luthringer, 1971; Rosen *et al.*, 1971; Chandra, 1972, 1974b, 1976b, 1977c; Neumann *et al.*, 1975; Bell *et al.*, 1976; Suskind *et al.*, 1977). There is no correlation between serum immunoglobulin levels and the degree of nutritional deficit, the presence of plasmacytoid lymphocytes in peripheral blood, or the prognosis. In some, especially those with a history of frequent gastrointestinal and respiratory infections, serum IgA is markedly increased (Keet and Thom, 1969; Chandra, 1972). Occasionally, serum IgG, and rarely IgM and IgA may be low, particularly in those who have the onset of nutritional deficiency in fetal or early postnatal life (Aref *et al.*, 1970; McFarlane *et al.*, 1970; Chandra, 1972, 1977c). After appropriate therapy with diet and antibiotics, the immunoglobulin levels tend to return to the normal range within a few days. Nutritional deficiency may reduce the synthesis and levels of IgG and perhaps of other immunoglobulins but, given the antigenic challenge of infection, there is accelerated production. In malnourished infected children, particularly those with gastrointestinal symptoms, there may be some loss of immunoglobulins as part of protein-losing gastroenteropathy. For example, measles infection is known to cause protein loss in the gut, and the deterioration in nutrition brought about by this

Table 6.1
Serum Immunoglobulin Levels[a]

Group	IgG (mg%)	IgA (mg%)	IgM (mg%)	IgD (mg%)	IgE (U/ml)
Energy–protein undernutrition					
With gross infection	2360 ± 654[c]	268 ± 56	188 ± 43	16.5 ± 7.1	360 ± 110
With demonstrable parasites[b]	2145 ± 721	309 ± 79	244 ± 57	29.3 ± 8.9	2865 ± 581
Without obvious infection or infestation	625 ± 207[d]	87 ± 23	81 ± 25	3.1 ± 1.9	427 ± 129
Well-nourished					
With gross infection	2895 ± 256[e]	220 ± 38	159 ± 33	11.1 ± 5.3	56 ± 37
With demonstrable parasites	1435 ± 216	180 ± 53	136 ± 47	7.6 ± 2.9	2410 ± 736
Without obvious infection or infestation	1080 ± 192[f]	110 ± 29	88 ± 21	1.9 ± 0.9	36 ± 21

[a] Values are expressed as mean ± standard deviation.
[b] *Ascaris* and/or hookworm.
[c] Plasma half-life of IgG = 9.6 days.
[d] Plasma half-life of IgG = 31.3 days.
[e] Plasma half-life of IgG = 11.1 days.
[f] Plasma half-life of IgG = 19.5 days.

process also contributes to immunosuppression produced by this disease.

A very wide range of values for serum IgD has been reported in healthy children. Thus the interpretation of serum IgD concentration in malnutrition is difficult, compounded by the lack of knowledge of the biological role of this immunoglobulin. Serum IgD is detectable by agar gel immunodiffusion in a higher proportion of undernourished children and the mean levels are higher than in well-nourished controls (Johannsson *et al.*, 1968; Suskind *et al.*, 1977; Chandra, 1978a) (Table 6.1).

Some children with kwashiorkor, especially those with infection, may show extremely high IgD concentration, as may some apparently healthy children. Serum IgD was increased in Gambian children but not in adults (Rowe *et al.*, 1968) and did not correlate with the presence of malaria or other infections (McGregor *et al.*, 1970).

In healthy, well-nourished children, serum IgE concentration is extremely low, often less than 10 units/ml requiring sensitive radioimmunoassay methods to detect it. Many exogenous agents as well as inherited factors influence IgE levels. In malnourished groups, serum IgE is significantly elevated (Table 6.1) (Johannsson *et al.*, 1968; Chandra 1977b, 1977f, 1978a). The increase is more pronounced in individuals with demonstrable parasitic infestations, most particularly *Ascaris lumbricoides*. However, undernutrition is so intimately intertwined with the occurrence of infections and infestations that it is almost impossible to define the effect of malnutrition *per se* on IgE synthesis. Invasive parasites such as roundworm and hookworm profoundly stimulate IgE (Chandra, 1977f). It has been suggested that IgE may be a protective factor in eliminating or reducing the number of parasites, and in reinfection immunity. The second contributing factor for elevation of IgE may be depressed T cell function. In other human and experimental states with reduced cell-mediated immunity, high levels of IgE have been reported (Kikkawa *et al.*, 1973). Interestingly, sera containing high concentrations of IgE inhibit *in vitro* lymphocyte response and sheep erythrocyte rosette formation by T cells (Chandra, 1977c, 1977d, 1977f, 1978a). Abbassy *et al.* (1974b) found that undernourished children did not develop immediate hypersensitivity reactions at the skin sites where reagin-containing serum was injected and then scratch-tested with the appropriate antigen. The failure to demonstrate passively transferred immediate hypersensitivity was ascribed to impaired release of chemical mediators of inflammation in protein–calorie undernutrition and/or nonresponsiveness of dermal tissue to these mediators.

In low-birth-weight infants, cord blood concentration of IgG is low and there is a significant correlation between serum IgE

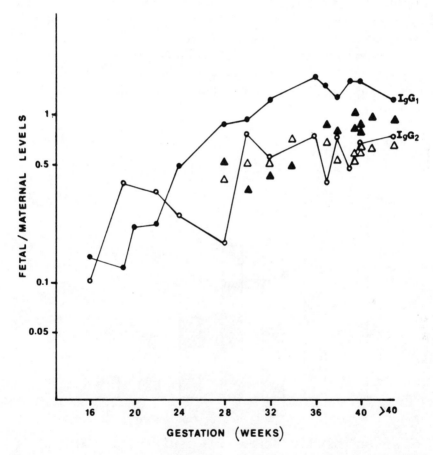

Figure 6.14. Ratio of fetal and maternal levels of IgG subclasses in paired samples in normal pregnancy. Values found in 12 infants with intrauterine growth retardation are also shown. IgG_1 = ● ▲, IgG_2 = ○ △. From Chandra, 1975b.

and birth weight (Chandra *et al.*, 1970; Chandra, 1975c). This may be the result of preterm delivery so that sufficient time is not available for acquisition of IgG from the mother, or in the case of small-for-gestation infants it may be due to placental dysfunction which may interfere with optimal transfer of IgG across the placenta (Chandra, 1975b). The transport of IgG_1 is affected more than that of IgG_2 (Fig. 6.14). Specific antibodies, such as tetanus antitoxin, are present in lower concentration in the sera of low-birth-weight infants (Fig. 6.15). The fetal/maternal ratio of tetanus antitoxin vaires from 0.25 to 1.46, depending upon gestational age and birth weight (Chandra, 1975b). The physiological hypoimmunoglobulinemia G is more pronounced in such infants (Fig. 6.16) and may be associated with the clinical problem of frequent infections. It tends to ameliorate by the age of 1 year, although in some infants low IgG levels may persist well into the second year

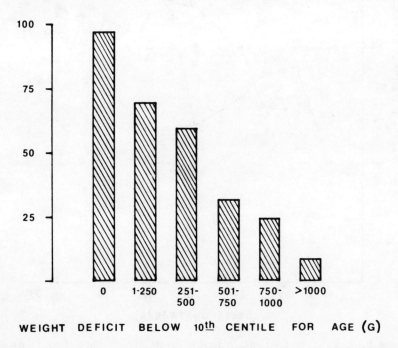

Figure 6.15. Percentage distribution of cord blood tetanus antitoxin levels above 0.01 unit/ml related to the degree of weight deficit in infants with intrauterine growth retardation. From Chandra, 1975b.

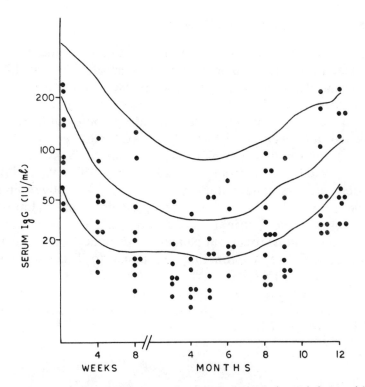

Figure 6.16. Longitudinal follow-up of serum IgG levels in 10 infants with fetal growth retardation, plotted on mean and range of values for healthy children. Between 3 and 6 months of age, some infants suffered from bacterial infections at a time when physiological hypoimmunoglobulinemia was pronounced. From Chandra, 1975c. Copyright 1975, American Medical Association.

of life (Aref et al., 1970). Subsequently, the antigenic stimulus of frequent infections increases the levels of all serum immunoglobulins including IgG, so that adult concentrations are reached by 2–5 years of age (Chandra and Ghai, 1972).

There is little information on the metabolism of immunoglobulins in malnutrition. Preliminary data suggest that the plasma half-life of radiolabeled IgG is decreased in those with high IgG concentration and is prolonged in those with hypoimmunoglobulinemia (Table 6.1). The latter parallels the findings in primary immunodeficiency states involving immunoglobulin synthesis.

6.3.3. Serum Antibody Response

Antibody titer following natural infection or immunization is one measure of humoral immunity. Levels of isohemagglutinins are normal. The data on antibody response in undernutrition are summarized in Table 6.2. In man, serum antibody response to most antigens given in adequate amounts is comparable in undernourished and well-nourished groups (Chandra et al., 1976b). Many of the studies in which impaired antibody synthesis has been reported suffer from lack of control of several critical variables such as concomitant infection, dose of antigen and the vehicle in which it is administered, severity of malnutrition, nature of specific nutrients which are deficient, simultaneous institution of nutritional supplements, presence of competitive microbes, and liver dysfunction. Response to T cell–dependent antigens may be affected more often (Fig. 6.17) because of the consistent impairment of cell-mediated immunity in malnutrition. Infection *per se* can act as an immunosuppressant irrespective of nutritional status (Chapter 5).

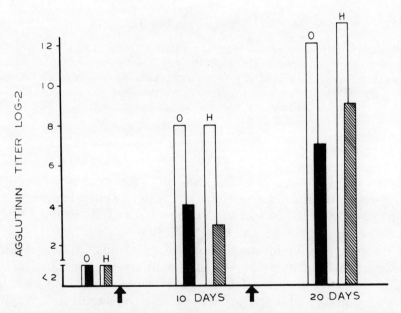

Figure 6.17. Agglutinin titer (geometric mean values) to *S. typhi* O and H antigens following primary and booster injections in healthy (open columns) and malnourished (blocked columns) children. From Chandra, 1972.

Table 6.2
Serum Antibody Response in Nutritional Deficiency[a]

Antigen	Group[b]	Antibody production[c] Normal	Antibody production[c] Reduced[d]
Bacterial			
Anthrax	A	Werkman (1923) Werkman et al. (1924a,b)	—
Brucella	A	Cooper et al. (1974)	Orr et al. (1931)
Cornybacterium kutscheri	A	Zucker et al. (1956)	Zucker et al. (1956)
Diphtheria toxoid	H	Balch (1950) Chandra (1977i)	Olarte et al. (1956)
	A	Pruzansky & Axelrod (1955) Klimentova & Frjazinova (1965)	Arkwright & Zilva (1924) Bieling (1925) Pruzansky & Axelrod (1955) Axelrod et al. (1961) Krishnan et al. (1974)
Escherichia coli	A	—	Orr et al. (1931)
Klebsiella	A	—	Benditt et al. (1949) Wissler et al. (1957)
Pasteurella tularensis	H	—	Morey & Spies (1942)
	A	—	Berry et al. (1945)
Pneumococcus	H	Neumann et al. (1975)	—
	A	Werkman (1923) Werkman et al. (1924a,b)	Wissler (1947)
Salmonella (killed bacilli)	H	Hodges et al. (1962a) Chandra (1975d) Chandra et al. (1977d)	Wohl et al. (1949) Meyer et al. (1955) Hodges et al. (1962b) Reddy & Srikantia (1964) Chandra (1972) Suskind et al. (1977)
	A	Werkman (1923) Werkman et al. (1924a,b) Blackberg (1927–28)	Orr et al. (1931) Greene (1933) Cannon et al. (1943) Meyer et al. (1956) Wissler et al. (1957) Harmon et al. (1963) Panda & Combs (1963)
Salmonella (flagellin)	H	—	Mathews et al. (1972, 1974)
Tetanus toxoid	H	Chandra (1972, 1975d) Kielman et al. (1976) Chandra et al. (1977d)	Hodges et al. (1962a, 1962b)
	A	Lopez et al. (1972)	Tashmukhamedov (1965) Krishnan et al. (1974)

Continued

Table 6.2 *(Continued)*

Antigen	Group[b]	Antibody production[c]	
		Normal	Reduced[d]
Viral			
Bacteriophage ϕX 174	A	Lopez et al. (1972)	—
Hepatitis	H	—	Chandra (1977i)
Influenza	H	—	Jose et al. (1970)
	A	Axelrod & Hopper (1960) Underdahl & Young (1956)	Axelrod & Hopper (1960)
Measles[e]	H	Chandra (1975f) Chandra et al. (1977d)	Mata & Faulk (1973)
Poliomyelitis[e]	H	Hodges et al. (1962c) Brown & Katz (1965) Chandra (1975f)	Chandra (1975c)
Tobacco mosaic	H	—	Gell (1948)
Western equine encephalomyelitis	A	—	Brown & Katz (1966)
Yellow fever	H	—	Ruckman (1946)
Miscellaneous			
Alloantigens	A	Malavé & Layrisse (1976)	Malavé & Layrisse (1976)
Ascaris	A	—	Leukskaja (1964)
Heterologous proteins	A	—	Gemeroy & Koffler (1949) Stavitsky (1957)
Heterologous erythrocytes	H	—	Gell (1948)
	A	Werkman (1923) Werkman et al. (1924a,b) Stoerk & Eisen (1946) Stoerk et al. (1947) Lopez et al. (1972)	Greene (1933) Stoerk & Eisen (1946) Wissler et al. (1946, 1957) Stoerk et al. 1947) Axelrod et al. (1947) Agnew & Cook (1949) Ludovici et al. (1949) Ludovici & Axelrod (1951a,b) Benditt et al. (1949) Kenny et al. (1968, 1970) Cooper et al. (1970, 1974) Mathur et al. (1972) Nalder et al. (1972) Aschkenasy (1973) Chandra et al. (1973, 1977d) McFarlane & Hamid (1973) Gebhardt & Newberne (1974) Krishnan et al. (1974) Chandra (1975g) Olusi & McFarlane (1976)

Table 6.2 *(Continued)*

Antigen	Group[b]	Antibody production[c] Normal	Reduced[d]
Keyhole limpet hemocyanin	H	Neumann et al. (1975) Chandra (1977i)	—
Rickettsia	A	Wertman & Sarandria (1951) Klimentova & Frjazinova (1965)	Wertman & Sarandria (1951) Klimentova & Frjazinova (1965)
Tumor	A	—	Jose & Good (1971, 1973)

[a] Modified from Chandra et al. (1976b).
[b] H = human subjects; A = laboratory animals.
[c] In some instances, antibody titer was affected by the deficiency of some nutrients but not by others. Or antibody response to some antigens was normal and to others it was reduced.
[d] Many studies in man have not carefully distinguished the immunodepressive effect of nutritional deficiency from that of concomitant infection.
[e] Secretory antibody response is reduced in energy-protein undernutrition (Chandra, 1975f), and in iron deficiency (Chandra et al., 1977d).

In some studies, antibody responses have been compared in groups of malnourished children given different levels of dietary nutrients, most commonly protein. Those on higher calorie-protein intake have higher antibody response.

In low-birth-weight infants, the capacity to form antibodies may be slightly impaired (Chandra, 1975c). For example, serum antibody titer, following Sabin polio virus vaccine, is lower in such infants than in healthy full-term controls (Fig. 6.18).

Geographic variations in the response of young children to oral polio vaccine have been recognized. However, factors other than nutrition may be important here. Sabin (1959) suggested that interference between concurrent infection with enteroviruses and vaccine virus may explain the low serum antibody conversion rate and delayed excretion of polio virus. Ghosh et al. (1970) found an extremely low incidence of rise in serum antibody. However, the potency of the vaccine at the time of administration was not checked and the nutritional status of the vaccinees was not mentioned. In a study from South India, Jacob John (1975) confirmed that seroconversion after oral polio virus vaccine was

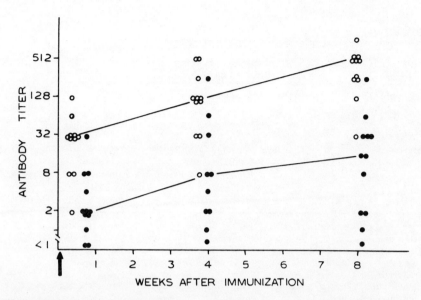

Figure 6.18. Neutralizing antibody reciprocal titers to poliovirus Type I in full-term healthy infants (○) and in infants with fetal growth retardation (●). The mean titer was lower in the latter group. From Chandra, 1975c. Copyright 1975, American Medical Association.

low. In the large majority, there was a good correlation between the absence or presence of vaccine virus excretion and negative or positive seroconversion, suggesting that impaired antibody response was mainly due to lack of virus "take." In a few individuals, seroconversion occurred in the absence of demonstrable fecal excretion of vaccine virus. This may reflect inactivation of virus by secretory antibody, nonuniform distribution, or intermittent excretion of virus in feces and inadequacy of sampling. In another group of vaccinees, the opposite phenomenon of virus excretion without serum antibody response was observed. None of these studies reported the nutritional status of the children who were vaccinated, nor was secretory antibody examined. Antibody response improved and was adequate when polio virus vaccine was administered five times (Jacob John, 1976).

In man, it is impossible to dissect out the effect of deficiency of a single nutrient on antibody-forming capacity. In earlier studies implicating specific deficiencies such as that of pyridoxine and pantothenic acid in reduced capacity to synthesize antibodies, the concomitant presence and influence of deficiencies of other nutrients was not clearly ruled out. Iron deficiency does not alter antibody response to a variety of antigens (Chandra 1975d, 1976a; Chandra *et al.*, 1977d).

In some undernourished subjects, the administration of an antigen may elicit a secondary booster response, suggesting previous exposure and memory, even though the preimmunization titer may be below the currently detectable levels. Small amounts of preexisting antibody may facilitate further antibody synthesis by one of many mechanisms: macrophage uptake and processing of antigen, presence and number of memory cells, etc.

6.3.4. Secretory Antibody Response

The importance of the secretory immune responses in protection from mucosal invaders is recognized. The unique characteristics of secretory antibodies have been recently defined (Hanson and Brandtzaeg, 1973).

In malnourished children, the concentration of secretory IgA (sIgA) in nasopharyngeal secretions is low (Chandra, 1975f; Sirisinha *et al.*, 1975). Since levels of total proteins and albumin were only marginally affected, reduction in sIgA could not be explained on the basis of a general impairment of protein synthesis nor by enzymatic proteolysis. The possibility of sIgA being reabsorbed through the mucous membrane was excluded by the absence of reaction between the serum and antiserum to secretory component. The reduction of sIgA may be the result of selective depression of IgA synthesis in submucosae or of secretory component production by epithelial cells (Chandra, 1977g). The mucosal epithelial cells, particularly in the gut, are severely depleted in number in undernourished subjects. An additional pathogenetic mechanism may be the reduction in the number and/or function of recirculating small lymphocytes which modulate IgA synthesis in submucosae (Chandra, 1977i). Gut-associated lymphoid aggregates, such as the tonsils, Peyer's patches, and lymphoid follicles in the appendix, are small in size in persons with nutritional deficiency.

In contrast, Bell et al. (1976) found significantly lower levels of albumin and higher levels of immunoglobulins in the intestine, the latter being attributed to the antigenic challenge provided by frequent severe infections. The ratio of immunoglobulins to albumin levels suggested that some of the gut luminal immunoglobulins may have been derived from plasma.

It has been postulated that immunoglobulins in intestinal secretions may be broken down into small low molecular weight fragments which would diffuse more rapidly through gel, giving larger precipitin rings and fallaciously high estimates of immunoglobulin concentrations. However, studies by Bell et al. (1976) looking at sepharose column fractionation of I^{125}-labeled duodenal fluids of Indonesian malnourished and infected children showed that more than 90% of IgA and IgG were present in molecular weight sizes of 350,000 and 150,000 respectively. Thus it is not likely that small fragments of immunoglobulin molecules, if present, alter estimated results.

Specific antibody response in mucosal secretions in nutritional deficiency has been examined in only one study (Chandra, 1975f). In a group of children with moderate or severe energy-protein undernutrition, live attenuated polio virus vaccine and measles virus vaccine were administered. Seroconversion was achieved in all the children but nasopharyngeal IgA antibody to the viral antigens was either undetectable or present in very low titer (Fig. 6.19). The mean secretory antibody level was significantly lower in the malnourished group compared with matched well-nourished children. However, a low titer of antibodies of IgG and IgM class was present, an observation similar to that made in IgA deficiency.

An additional characteristic of gut immunity, namely the lack of an anamnestic or secondary component (Newcomb et al., 1969), may be pertinent to the above findings. In malnutrition, antibody response to primary immunization challenge may be inadequate, but secondary booster titer is often normal. In the respiratory and gastrointestinal secretions, protection from infection and systemic spread requires an adequate "primary" response each time. This may be reduced in malnutrition, as suggested by the preliminary observations (Chandra, 1975f).

IMMUNOCOMPETENCE IN UNDERNUTRITION

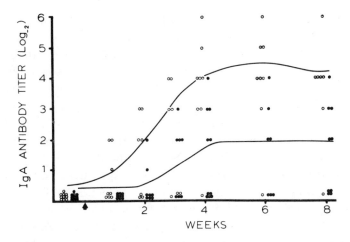

Figure 6.19. Titers of IgA antibody to measles antigen in nasopharyngeal secretions following immunization with a single dose of live attenuated measles vaccine. Healthy = ○, undernourished = ●. Almost one-half of children with energy-protein malnutrition did not have any detectable IgA antibody. Serum antibody response, not shown here, was comparable in the two groups. From Chandra, 1975f.

There is no published data so far on the mucosal cell-mediated responses in malnutrition.

The clinical and biologic significance of reduced secretory antibody response may touch upon several known facets of undernutrition. Impaired secretory immunity may contribute to an increased frequency of infections often seen in association with nutritional deficiency. The small amounts and reduced activity of luminal sIgA may fail to prevent mucosal binding of bacteria and enterotoxin, a prerequisite of functional and metabolic abnormalities caused by several enteropathogenic organisms. Systemic spread may also occur more easily because of the reduced efficiency of the mucosal barrier in checking pathogenic organisms from penetrating respiratory and gastrointestinal epithelia. Other macromolecules, such as dietary proteins, pollen, etc., which are normally excluded by the mucosal barrier may also get across the mucous membrane. Food antibodies of IgG and IgA classes are frequently found in high titer in malnourished children

(Chandra, 1975e). However, complement activation has not been observed and no adverse effects were seen after ingestion of the dietary proteins. It is not known if the frequency occurrence of other immunopathologic diseases known to be associated with defective mucosal immunity such as allergy, autoimmunity, and neoplasia, is increased in malnourished groups. Individuals with low or absent IgA in saliva and other secretions have a high incidence of atopy and collagen disorders. It is possible that chronic persistent stimulation of lymphoid tissue by dietary antigenic fragments absorbed freely through the gastrointestinal mucous membrane may act together with other factors such as genetic susceptibility and oncogenic viruses to produce malignant change. This may also occur in the event of excessive intake and consequent heavy antigenic load in the gut. A recent analysis of geographic variations in prevalence of disease revealed a positive correlation between consumption of animal protein and lymphoma mortality (Cunningham, 1976). At the same time, the effect of diet on immunocompetence may alter, increase, or decrease host resistance to tumor growth (Chapter 8).

6.4. COMPLEMENT SYSTEM

The role of the complement system in amplification of the immune response, including opsonization, immune adherence, phagocytosis, leukocyte chemotaxis, and viral neutralization is established (Johnston and Stroud, 1977). With malnutrition there is an increased incidence of infections which tend to be more severe and prolonged, and subjects are slower to recover from gram-negative organisms which often plague the undernourished. In defense against these organisms, the classical and alternative pathways of complement activation play an important role. Also, it is known that congenital deficiencies of C3 and other complement components may be associated with frequent pyogenic infections, collagen disorders, and diseases of autoimmune etiology. From these considerations, studies of the quantitation of the complement components and their functional activity in malnourished children emanated.

Total hemolytic activity of the serum is a measure of the

functional integrity of the complement cascade. Sera of undernourished children showed a reduction in total hemolytic complement (Table 6.3) (Chandra, 1975a). Smythe et al. (1971) found reduced complement activity in malnourished South African children. In many patients, the Coombs direct antiglobulin test was positive due to the presence of C4 (and occasionally other complement components) and immunoglobulin on the surface of red cells which may contribute to a shortened erythrocyte survival. However, there was no correlation between reduced hemolytic serum complement and the presence or absence of C4 on erythrocytes. It is possible that depressed thymus function in malnutrition facilitates autoimmunization. Chandra (1972) reported low concentrations of complement C3 in Indian infants who had reduced body weight in relation to the expected for age and height. It was suggested that low synthetic rate of complement components in the liver, gut, and macrophages, as well as increased consumption of complement proteins in antigen–antibody reactions, result in reduced complement activity. C3 levels are low in hepatitis (Chandra, 1970). Sirisinha et al. (1973) studied the complete complement profile of 20 malnourished Thai children, some with kwashiorkor and others with marasmus. Serum concentrations of C1q, C1s, C3, C5, C6, C8, C9, and C3 proactivator were markedly lower in the undernourished group than in

Table 6.3
Hemolytic Complement, C3, and Immunoconglutinin in Healthy and Malnourished Children, with or without Infection

	Well-nourished		Undernourished	
	No infection	With infection	No infection	With infection
No. of individuals studied	20	10	23	12
CH_{50} (units/ml)[a]	58 ± 13	105 ± 21	39 ± 15	28 ± 11
C3 (mg/100 ml)[a]	132 ± 18	245 ± 57	89 ± 23	57 ± 19
Altered C3 on immunoelectrophoresis (no. positive)	0	6	9	5
IK ($-\log$)[b]	1	5 (2–8)	3 (<1–5)	4 (<1–8)

[a] Mean ± standard deviation.
[b] Mean (and range).

normal children of the same age in the same geographical area (Fig. 6.20). Complement component C4 level was comparable in the two groups. After an initial week of stabilization, the patients were randomly allocated to 1 of 4 dietary regimens with caloric intake of 100 or 175 C/kg/day and protein intake of 1 or 4 g/kg/day. The number of children in each of these nutritional rehabilitation schemes was small (2–8) but the children who received the high-calorie high-protein diet tended to respond best in terms of recovery of levels of complement components. The posttherapy concentrations were above normal values (Fig. 6.20) possibly a consequence of accelerated synthesis and overproduction. Alternatively, this "rebound" elevation may be a response to residual infection, since complement activity is often increased in well-nourished subjects with infection (Table 6.3). On the other hand, infection in association with nutritional deficiency depresses complement activity and C3 concentration. There is a significant correlation between levels of C3 and duration of infectious illness, the complement level decreasing or increasing depending upon the nutritional status (Fig. 6.21).

In 35 infants and young children with primary energy–protein undernutrition studied in New Delhi, assays of complement and immunoconglutinin were done and the results analyzed in relation to coexisting infection, complement activation *in vivo*, and the effect of dietary rehabilitation (Chandra, 1975a). There was a wide scatter of values in individual children, but the mean hemolytic activity and concentration of C3 in malnourished children was lower than that of healthy matched controls, the reduction being more pronounced in those with overt infection. In addition, 42% of serum samples showed anticomplementary activity, which could be due to one or more of the following causes: presence of antigen–antibody complexes, endotoxemia (Oberle *et al.*, 1974), macroglobulin, and inhibitors of C1 esterase, C3, and C6. In some serum samples, several tests for detection of circulating immune complexes were positive (Table 6.3) (Chandra, 1977b), but the nature of the antigen(s) involved was not established. Reduced levels of total complement hemolytic activity and C3 in malnutrition have been recently confirmed by Suskind *et al.* (1976) and Neumann *et al.*(1977). Similar observations have been made in acute starvation (Palmblad *et al.*, 1977a).

Low levels of complement activity in nutritional deficiency

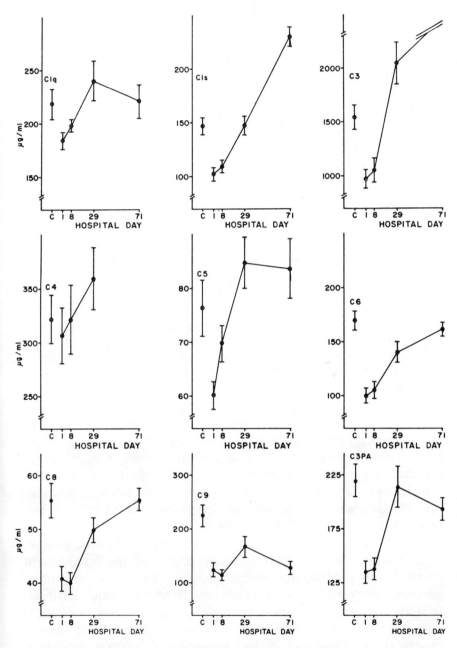

Figure 6.20. Serum concentrations of complement proteins in severe energy-protein undernutrition. Estimations were done on admission to hospital and sequentially for several weeks during which the children were getting nutritional supplements. From Sirisinha et al., 1973.

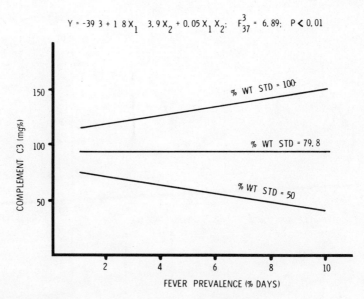

Figure 6.21. Serum C3 levels correlated with infection–morbidity indicated by the number of days of fever. Three weight categories in reference to the standard are shown. In well-nourished children, C3 concentration increases with infection. In the undernourished, C3 levels generally decrease. The child population examined is fully described elsewhere (Kielman et al., 1976). Courtesy of Dr. A. Kielman.

may reflect a general reduction in protein synthesis or mobilization of the limited synthetic ability for production of more urgently required antibodies directed against the invading pathogen. Lymphoid tissue, a site of complement synthesis, is severely involuted in states of nutritional deprivation, and cell division, regeneration, and proliferation are restricted. Additionally, liver involvement invariably seen in malnutrition may reduce hepatocytic production of C3 and other components of the complement system. The gut and the macrophage system synthesize some complement proteins and these tissues are altered in malnutrition. In infected undernourished children, evidence of complement consumption and activation *in vivo* was adduced from several observations (Chandra, 1975a): presence of electrophoretically altered complement C3 (Fig. 6.22), a reciprocal rise in titer of immunoconglutinin (Fig. 6.23) which is an antibody to converted C3 and C4, and presence of anticomplementary activity in serum.

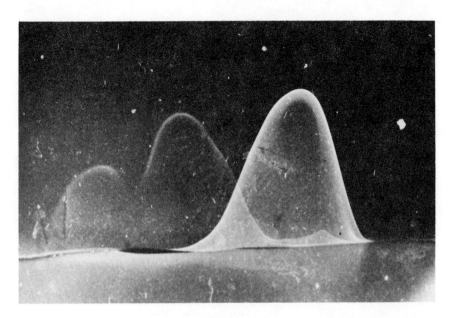

Figure 6.22. Two-dimensional cross-over immunoelectrophoresis into gel containing antiserum to C3. Two additional peaks were observed in some serum samples of undernourished children with infection. These peaks represent C3 conversion products often designated C3b and C3c. Cathode is to the right. From Chandra, 1977f.

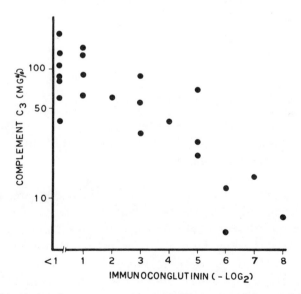

Figure 6.23. Correlation between complement C3 concentration and immunoconglutinin titer in energy–protein undernutrition. Children with low levels of C3 had raised titers of immunoconglutinin. From Chandra, 1975a.

Immunoconglutinin tends to appear with a lag period and persists much longer than other changes suggestive of complement activation.

The alterations in complement activity and levels of component proteins seen in malnourished subjects may have a contributory role in the susceptibility of such individuals to frequent life-threatening gram-negative septicemias.

6.5. PHAGOCYTES

The important role of the polymorphonuclear leukocytes (PMNs) and macrophages in primary defense against microbes and in the afferent limb of the immune response is well recognized. This is dramatized and confirmed by the range and magnitude of clinical manifestations, particularly life-threatening infections, seen in patients with congenital or acquired abnormalities of PMN function. The syndromes include neutropenia, chronic granulomatous disease, myeloperoxidase deficiency, glucose-6-phosphate dehydrogenase deficiency, Chediak-Higashi syndrome, familial lipochrome histiocytosis, lazy-leukocyte syndrome, extensive burns, and trauma. Many recent studies have investigated the production, mobilization, and phagocytic, metabolic, bactericidal aspects of phagocyte function in energy–protein undernutrition.

6.5.1. Number and Morphology

The total leukocyte count in nutritional deficiency is commonly increased or normal. Relative neutrophilia is frequent. Fulminant sepsis in malnourished patients can reduce the number of total white cells including PMNs.

The microscopic appearance of PMNs in undernutrition is largely conditioned by the occurrence of infection. In noninfected children with energy–protein undernutrition, morphologic or fine structural alterations have not been noted. However, concomitant subclinical or overt sepsis produces a shift to the left and toxic granulation (Chapter 5). There is an increase in rough-surfaced endoplasmic reticulum and the Golgi zones are prominent (Fig. 6.24a,b). Occasionally, Döhle bodies were observed. Myelin

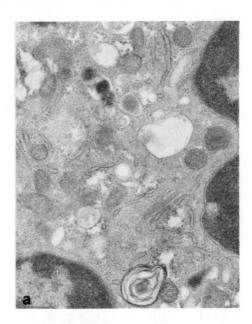

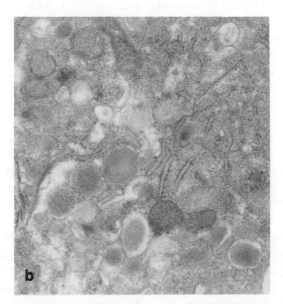

Figure 6.24. Fine structure of neutrophil from a kwashiorkor patient. (a) Well-developed rough-surfaced endoplasmic reticulum which is characteristic of Döhle bodies. × 32,000. (b) Prominent Golgi zone. × 22,000. From Schopfer and Douglas, 1976b.

bodies and cytoplasmic vacuoles were often encountered. These features are typical of immature PMNs (Bainton et al., 1971) and have been seen in association with PMN dysfunction in patients with extensive burns (Alexander and Wixson, 1970).

6.5.2. Mobilization

There are no published reports on the full profile of neutrophil kinetics in malnutrition. However, there is some data available on the mobilization of PMNs from the marginal and marrow pools of granulocytes. Injection of adrenaline constricts the spleen and within a few minutes releases neutrophils from other sequestered marginal sites. The increase in neutrophil count in the peripheral blood is a rough index of the marginal pool. This pool is unchanged in patients with malnutrition and infection, singly or in combination (Chandra et al., 1976a). The administration of *Pseudomonas* polysaccharide is also followed within a few hours by neutrophilia, which is an index of the marrow reserves. This is significantly reduced in malnutrition as well as in infection (Chandra et al., 1976a), the effects of the two factors being additive. It is possible that in nutritional deficiency frequently associated with infection, the marrow pool of granulocytes is being constantly discharged into the blood under the repeated stimulation of pathogens and their products. Circulating endotoxin is frequently detected by the Limulus assay in such patients. Other unknown factors promoting PMN release from the bone marrow may also be at work.

6.5.3. Chemotaxis and Inflammatory Response

The ability of the PMN to act as a primary defense barrier is dependent on its unique capacity to sense and reach the sites of bacterial invasion. The inflammatory response is critical to the host's ability to localize the infection and to provide amplification systems to deal with the pathogen. In malnourished children, the pattern of inflammatory response has been suspected to be altered. Gangrene rather than suppuration may take place in response to bacterial challenge.

In a preliminary report on inflammatory responses in a few malnourished children, Kumate (1969) found no differences between patients and controls. However, the observations were made only for a period of 6 hours. Subsequently, Freyre *et al.* (1973) used the Rebuck skin-window technique to evaluate the inflammatory response in 33 Peruvian children suffering from kwashiorkor or marasmus and 10 controls. In each subject, the presence of obvious infection was excluded. The total cellularity observed in the undernourished children was quantitatively similar to that of the control group, and the proportion of the two main types of inflammatory cells, PMNs and monocytes, was comparable in the two groups in the first observation period of 6 hours following scarification. In the subsequent 7–24 hour period, there was a reduction in the percentage of mononuclear leukocytes in the malnourished group, a pattern resembling the findings in neonates. In fact, at no time during the investigation period of the inflammatory cycle was a predominance of mononuclear leukocytes observed even in healthy children in contrast to findings in other studies. It is likely that this difference was related to the effect of the high altitude (7,000 feet) at which the Peruvian study was conducted. The morphologic and time-sequence alterations in the inflammatory response were confirmed in a group of undernourished infants in New Delhi (Chandra, 1977b) and in Thailand (Kulapongs *et al.*, 1977b). The reduced migration of mononuclear leukocytes to the site of challenge (physical, microbial antigen) may contribute to impaired cutaneous delayed hypersensitivity consistently observed in energy–protein undernutrition. In small-for-gestation low-birth-weight infants, the additive effects of neonatal handicap, nutritional wasting, and infection are observed.

The random mobility of PMNs obtained from children with energy–protein undernutrition is only marginally altered. Early chemotactic migration is reduced (Schopfer and Douglas, 1975) but the overall capacity for chemotaxis is preserved unless complicating infection occurs (Rosen *et al.*, 1975; Chandra *et al.*, 1976a). Sepsis depresses chemotactic movement of PMNs.

6.5.4. Metabolism

There is limited data on the metabolic activity of PMNs in nutritional deficiency. The relevance of increases in glycolytic and oxidative pathways during phagocytosis to particle uptake and destruction is established (Selvaraj and Sabarra, 1966). Morphological data indicated normal phagocytic activities of leukocytes in malnutrition (Balch and Spencer, 1954; Tejada *et al.*, 1964). Yoshida *et al.* (1968) reported a low endogenous level of lactate and decreased pyruvate kinase activity in leukocytes isolated from patients with energy-protein undernutrition, suggesting reduced glycolytic activity. Selvaraj and Bhat (1972a) confirmed this finding and observed a failure of increase in lactate production normally seen in response to particle challenge and uptake.

Enzymes and metabolic pathways pertinent to bactericidal activities of leukocytes have been assayed in patients suffering from energy-protein undernutrition. The resting level of the hexose monophosphate shunt was increased, possibly due to associated infection (Selvaraj and Bhat, 1972a). Granule-bound NADPH oxidase activity was low and failed to show the phagocytic stimulation normally seen in cells of healthy donors (Selvaraj and Bhat, 1972b). The deficiency of NADPH oxidase in malnutrition may be the direct result of protein deficiency or hypercortisolemia observed in nutritional deficiency (Rao *et al.*, 1968). Cortisol inhibits NADH oxidase of human leukocytes (Mandell *et al.*, 1970). There was little or no release of acid phosphatase from lysosomes during phagocytosis. There was a reduction in iodination as measured by ^{131}I incorporation into trichloroacetic acid-precipitable proteins (Schopfer and Douglas, 1975). These data suggest an impairment of factors involved in the intracellular myeloperoxidase hydrogen peroxide-iodide-mediated bacterial-killing processes. However, myeloperoxidase activity was reported to be normal in PMNs of kwashiorkor children (Avila *et al.*, 1973) but is reduced in iron deficiency (Higashi *et al.*, 1967). It is likely that altered glycolytic and oxidative metabolism of PMNs in malnutrition contributes to impaired microbicidal capacity of leukocytes. The metabolic abnormalities disappeared on correction of nutritional deficit. Activities of leukocytic glucose-6-phosphate dehydrogenase and 6-phosphogluconate dehydrogenase were normal (Selvaraj and Bhat, 1972b). Schopfer and Douglas (1976c) measured the activities of cytoplasmic enzymes including hexokinase, fructokinase, phosphoglycerate kinase, pyruvate kinase, glucose-6-phosphate dehydrogenase, and of membrane-bound enzymes including myeloperoxidase, NADPH- and NADH-oxidase, glutathione reductase, and anaerobic lactate production by resting and phagocytizing PMNs. No difference was observed between kwashiorkor and normal children. However, leukocyte iodination was reduced without any impairment of T4 degradation.

Nitroblue tetrazolium (NBT) test has been used as an index of intraneutrophilic metabolic activity. Using the qualitative test, Shousha and Kamel (1972) found that the percentage of cells with formazon compound formed from the tetrazolium dye was significantly less in Egyptian children with kwashiorkor than in the age-matched well-nourished control group. There was a direct correlation between the proportion of neutrophils with histochemical evidence of formazon and serum concentration of proteins, albumin, and hemoglobin, but not with total leukocyte counts. Surprisingly, this reduction in the NBT test was seen in spite of the presence of infections in the undernourished group. Another unexplained observation was the absence of marked differences between the results whether or not latex particles were used. It suggests that the PMNs of children in both groups were almost maximally stimulated *in vivo* and no further stimulation was achieved by the addition of latex. It also casts doubt on the freedom from infection assumed for the well-nourished control group. Moreover, it is unlikely that the ability of each PMN to reduce the dye is an all-or-none phenomenon. In spite of these drawbacks of the study, the observations indirectly indicate that the activity of NADH oxidase involved in the conversion of NBT is diminished. Kendall and Nolan (1972) also concluded that reduction of NBT in malnutrition was impaired. On the other hand, Altay *et al.* (1972) and Rosen *et al.* (1975) observed that NBT reduction remains unimpaired in nutritional deficiency and that infection, clinical or subclinical, was the critical determinant. The objections related to the use of the qualitative NBT also apply to these latter studies.

Quantitative NBT test (Baehner and Nathan, 1968) is more likely to provide discrimination between PMN function in well-nourished and undernourished groups if only a partial impairment of metabolic processes exists, as might be expected in the clinical state of nutritional deficiency that varies enormously from the asymptomatic to the preterminal. Chandra *et al.* (1977a) found that NBT reduction by the neutrophils of children with malnutrition was increased, an indication of stimulation *in vivo,* presumably by bacterial pathogens, their products or immune complexes. The maximum production of formazon on phagocytic stimulation was comparable in the well-nourished and undernourished groups; ΔO.D. was consequently reduced in the latter. It would be fallacious to interpret this as an impaired PMN metabolic function, since in the malnourished patients intracellular metabolic pathways are being constantly stimulated *in vivo* and only a limited further increase in this would achieve the maximum possible limit.

Wolfsdorf and Nolan (1974) also observed high resting levels of reduced NBT indicative of increased hexose monophosphate shunt in Rhodesian children with protein–calorie malnutrition. However, this study did not examine the effect of phagocytosis on NBT test.

6.5.5. Opsonization, Phagocytosis, and Bactericidal Capacity

In two simultaneous reports from India (Selvaraj and Bhat, 1972a; Seth and Chandra, 1972), a reduction in bacterial-killing capacity of PMNs was demonstrated. In a group of children with primary energy–protein undernutrition diagnosed on the basis of reduced dietary intake and weight deficit, the opsonic function of plasma and ingestion of *Staphylococcus aureus* was normal (Seth and Chandra, 1972). Nevertheless it is possible that opsonization is less than optimal and rate limiting in extravascular sites where opsonins are present in amounts considerably different than in serum. In undernourished children, the killing of phagocytized bacteria was slow and less efficient. Counts of viable intracellular bacteria were significantly higher at the end of 2 hours of culture. The differences in the microbicidal activities of PMNs from undernourished and control groups were maintained over the 12-hour observation period (Fig. 6.25). The severity of leukocytic defect was scattered over a wide range. Some values approached the figures reported in the inherited syndrome of chronic granuloma-

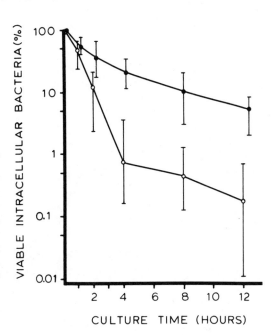

Figure 6.25. Intracellular bacterial killing capacity of polymorphonuclear leukocytes in patients with energy–protein undernutrition. Leukocytes and *Staphylococcus aureus* were mixed in the ratio of 1 : 5 and incubated at 37°C in the presence of 20% normal plasma. At various intervals viable intracellular bacteria were estimated by surface colony count on nutrient agar, and expressed as a percentage ratio of a similar count at 20 minutes. ○ = patients, ● = controls. From Chandra et al., 1977a.

tous disease. Selvaraj and Bhat (1972a) reported reduced capacity of PMNs to kill *E. coli*. The bactericidal activity was low or almost absent throughout the whole incubation period. Similar changes were observed in mononuclear phagocyte function (Chandra, 1977i). Nutritional recovery was associated with an improvement in bactericidal capacity of PMNs within a few weeks, often to well within the range for controls (Fig. 6.26).

The normal phagocytosis and reduced bactericidal ability of phagocytes in nutritional deficiency was confirmed in studies on 5 Ivory Coast children suffering from severe kwashiorkor. All the children had albumin levels below 2 g/100 ml. The viable intracellular bacterial count was comparable in controls and patients at 30 minutes but much higher in the latter at 60 minutes. It is unlikely that significant differences can be picked up in the early stages of cell–bacteria incubation, even in the gross defect associated with chronic granulomatous disease. The endocytosis of polystyrene particles and antibody-coated erythrocytes was adequate (Douglas and Schopfer, 1974). Keusch et al. (1977b) studied malnourished Guatemalan children and confirmed impaired killing of

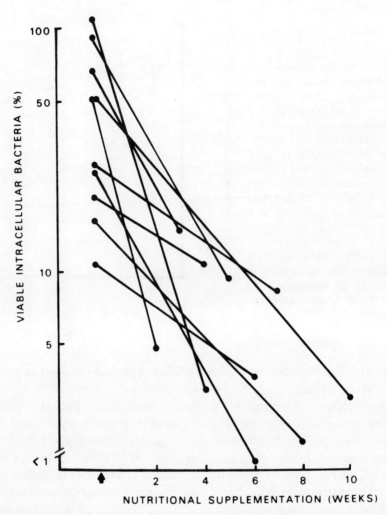

Figure 6.26. Impaired bactericidal capacity in undernourished children improved significantly after a short period of nutritional therapy. The majority of posttreatment values are within the range of results obtained in well-nourished controls. Experimental details as in Fig. 6.25. The viable intracellular bacteria were counted in aliquots obtained at 20 min and 140 min.

phagocytized bacteria. In addition, both resting and phagocytizing values of $^{14}CO_2$ production from 1-^{14}C-glucose were reduced but $^{14}CO_2$ release increased severalfold during phagocytosis, unlike results obtained in chronic granulomatous disease.

Reduced microbicidal activity of PMNs has also been observed in acute starvation (Palmblad, 1976) and in anorexia nervosa (Gotch *et al.*, 1975; Palmblad *et al.* 1977b).

Infection is an almost invariable complication of untreated severe malnutrition. It is likely that infection itself exerts a depressive influence on PMN function (Chapter 5). Chandra *et al.* (1977a) examined groups of well-nourished and undernourished children with and without obvious infection. Opsonization and phagocytosis were normal in all groups but bacterial-killing capacity was impaired in nutritional deficiency and in infection. In children who were undernourished and infected, the abnormalities of PMN function were additive. Leukocyte function could be improved considerably with antibiotic therapy for 5-7 days and restored completely back to normal by nutritional recovery within a few weeks (Fig. 6.27). McCall *et al.* (1971) and Rosen *et al.* (1975) showed that bacterial-killing capacity of neutrophils, particularly toxic granulocytes, of patients with sepsis is reduced.

The influence of specific nutrients on PMN function is not clear. It is possible that several dietary factors affect various facets of granulocyte physiology. Hypoglycemia depresses function of reticuloendothelial system (Buchanan and Filkins, 1976). Iron deficiency causes a reversible defect in microbicidal capacity and metabolic function evaluated by the quantitative NBT test (Fig. 6.13) (Chandra, 1973a, 1973b, 1975d, 1976a; Chandra *et al.*, 1977d; MacDougall *et al.*, 1975). Iron repletion brought PMN function back to normal. Contrarily, Kulapongs *et al.* (1974) studied 8 Thai children with severe iron–deficiency anemia and could find reduced bacterial-killing capacity of PMNs in only one patient. The differences in the results obtained in various studies are not easily explained but may well be due to differences in methods (for example, phagocyte/bacteria ratio, distinguishing leukocyte function from serum factors, range of microbes tested and their metabolic needs), presence of deficiencies of other nutrients, concomitant infection, etc.

It is interesting that overnutrition and obesity also reduce bacterial-killing capacity of granulocytes (Fig. 6.28). In a group of obese individuals, opsonization and ingestion were unchanged but intracellular bacterial inactivation was diminished. The mechanisms involved are not known. Similar findings have been recently reported by Palmblad (1977). The relationship of blood lipid concentrations to PMN function needs to be determined.

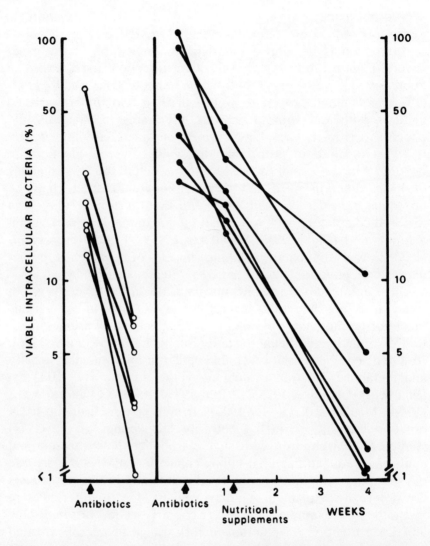

Figure 6.27. Influence of infection on bactericidal capacity of polymorphonuclear leukocytes in well-nourished (○) and undernourished (●) individuals, and effect of therapy with antibiotics and dietary supplements. Experimental details as in Fig. 6.25. Infection was associated with impaired bactericidal function of leukocytes. Improvement was noted within 1 week of antibiotic therapy and control of infection. In the undernourished group, nutritional therapy reversed the remainder of leukocyte dysfunction.

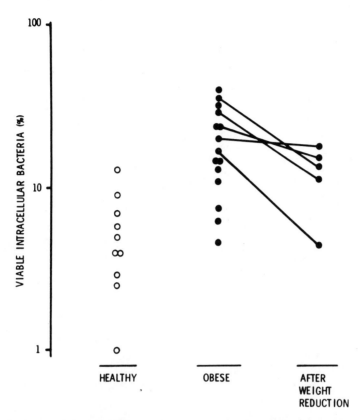

Figure 6.28. Bactericidal capacity of PMNs in obesity, defined on the basis of overweight and increased skin-fold thickness. In some individuals, the test was repeated after weight reduction.

6.6. LYSOZYME

Lysozyme (muramidase) is a small molecular weight (15,000 daltons) basic protein which lyses the cell-wall polysaccharides of many gram-negative bacteria, particularly those which possess an accessible β-1,4-glycosidic bond between N-acetylmuramic acid and N-acetylglucosamine. Other bacteria become susceptible to the action of lysozyme by pretreatment with antibody and complement, glycine, chelating agents, change in pH, ascorbic acid, and hydrogen peroxide. Lysozyme is present in high con-

Table 6.4
Lysozyme Activity[a]

Groups	Neutrophils (mg/10^6 cells)	Plasma (μg/ml)	Plasma/ neutrophil ratio	Urinary clearance[b] ($\times 10^3$)
Healthy	3.32	3.07	0.92	5.30
Well-nourished, infected	1.71	8.29	4.79	
Undernourished	2.15	1.63	0.76	0.73
Undernourished, infected	1.30	3.18	2.44	

[a] Adapted from Chandra et al. (1977b).
[b] Expressed relative to creatinine clearance.

centration in PMNs and monocytes. It is also present in various body fluids including serum, and in secretions, particularly tears and saliva. It is filtered through the glomerulus and reabsorbed mainly by the proximal convoluted tubule. The plasma level of lysozyme is governed by intracellular synthesis, release, degradation, passage into various secretions, and excretion through the urine, feces, tears, etc.

There is limited information on lysozyme concentration and activity in nutritional deficiency. Undernutrition reduces lysozyme activity in the plasma and in neutrophils (Chandra et al., 1977b). The presence of complicating infection increases plasma lysozyme, probably a result of its extrusion from PMNs and monocytes which have lower intracellular concentration (Table 6.4). The plasma/neutrophil lysozyme ratio is comparable in noninfected groups, whereas undernourished patients with infection had a significantly lower plasma/neutrophil ratio than the well-nourished infected controls. The urinary clearance of lysozyme was 3- to 12-fold higher in malnourished children. Mohanram et al. (1974) showed that the activity of lysozyme in leukocytes was significantly reduced in children with kwashiorkor and that dietary therapy restored the enzyme levels to normal.

The significance of alterations in lysozyme concentration inside the leukocytes or in the plasma of undernourished groups is not clear, but a lowered defense capacity on mucosal surfaces may result.

6.7. IRON-BINDING PROTEINS

The bacteriostatic effect of transferrin discovered by Schade and Caroline (1946) has been repeatedly confirmed. The addition of iron *in vitro* enhances growth of many bacterial and fungal species. Sera with high unsaturated iron-binding capacity impair multiplication of yeast whereas the iron-saturated serum from patients with thalassemia major promotes fungal growth. The elegant observations of the Bullens and their colleagues have proven that the bacteriostatic effect of human breast milk is due mainly to the large quantities of lactoferrin and transferrin contained in it (Bullen *et al.*, 1972). These iron-sequestering proteins, acting in concert with specific antibody, strongly inhibited the growth of bacteria (Fig. 6.29), an effect that was abolished by saturating the lactoferrin. In several clinical situations, iron availability is one of the major determinants of infection (Chandra *et al.*, 1977d).

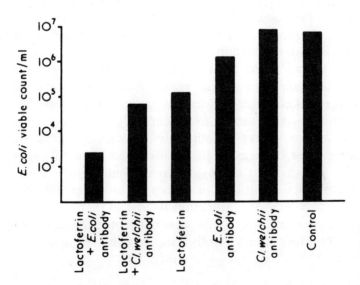

Figure 6.29. Effect of purified lactoferrin and antibody on growth of *E. coli* 0111 in medium 199. pH = 7.4; pO_2 = 80–90 mm Hg; inoculum = 2×10^2 organisms. Incubation for 5 h 30 min. Antisera were previously heated to 56°C for 30 min. From Bullen *et al.*, 1972.

Serum transferrin concentration is frequently reduced in energy–protein undernutrition and correlates significantly with the severity of nutritional deficiency. In kwashiorkor, transferrin level is low and survival correlates directly with the concentration of this iron-binding protein (Antia *et al.*, 1968). Death from overwhelming sepsis is sometimes a complication of iron administration in kwashiorkor before nutrition supplements have improved the general health status of the patient. In patients with chronic pyelonephritis, injection of iron sorbital citrate is associated with increased urinary output of white blood cells.

6.8. TISSUE INTEGRITY

Damage to tissues results in increased local susceptibility to infection. Horwitt (1955) reviewed the pathologic tissue changes brought about by nutritional deficiency. These include alterations in intercellular substance, reduction of mucus secretion, increased permeability of mucosal surfaces, accumulation of cellular debris and mucus, resulting in favorable conditions for growth of microorganisms, keratinization and metaplasia of epithelial surfaces, loss of ciliated epithelium of the respiratory tract, nutritional edema with increased fluid in tissues, reduced fibroblastic response, and impaired tissue replacement and repair. Present knowledge makes it difficult to evaluate objectively the contribution of such tissue changes to susceptibility or severity of infections. It is not known whether deficiencies of some nutrients produce greater changes in the tissues than those of others. For example, avitaminosis A results in epithelial metaplasia in the respiratory mucosa but lack of other vitamins does not cause such morphologic alterations. Infection itself can damage tissues for a variable period of time. For example, measles and whooping cough alter respiratory epithelium making it more prone to bacterial invaders and secondary infection.

6.9. INTERFERON

The interferon system is a mechanism of cellular immunity shared by man with most animals. Interferons, protein substance(s) produced under the stimulus of viruses and other inducers, confer antiviral resistance upon normally susceptible cells usually of the same species. The mechanisms of interferon action are not known. It may act through uncoupling of oxidative phosphorylation, thereby reducing the amount of ATP available for viral replication within the cell. Host protein synthesis is cut off, viral-specific polysomes and viral-coded proteins are not produced, and all subsequent events in viral biosynthesis do not occur (Grossberg, 1972).

The effect of nutritional deficiencies on the production and release of interferon is largely unknown. Schlesinger *et al.* (1977) studied the production of interferon by leukocytes of nine marasmic infants. Interferon synthesis was induced *in vitro* in lymphocyte cultures stimulated with Newcastle virus and assayed by determining lytic-plaque formation on monolayers of Vero cells challenged with bovine vesicular stomatitis virus. Interferon production was significantly less in leukocyte cultures of marasmic infants when compared to matched healthy controls. These preliminary observations of impaired synthesis of interferon, if confirmed in other studies, may partly explain the enhanced severity of some viral diseases such as measles and herpes simplex in malnourished individuals.

6.10. OTHER FACTORS

6.10.1. Gut Microflora

Alteration of intestinal flora in terms of both quality and quantity can contribute to incidence of gastroenteritis and malabsorption in malnutrition. Nutritional deficiency may be associated with the presence of large numbers of pathogenic bacteria in the upper small gut, which is normally sterile. Work on experimental animals demonstrates a frequent alteration of gut bacteria when acute or subacute starvation is induced. It is possible that a similar phenomenon takes place in weanling diarrhea, in which a synergism between infection and undernutrition is postulated to be the underlying mechanism. It has been postulated that bacteria in the jejunum can contribute to nutritional deficiency by deaminating dietary protein, bacterial degradation of amino acids, production of unconjugated bile salts, and fat malabsorption. Changes in gut microflora and fecal colonization have been documented in human energy–protein undernutrition (Gracey and Stone, 1972; Mata *et al.*, 1972; Heyworth and Brown, 1975). On the other hand, James *et al.* (1972) did not find any correlation between the numbers and types of jejunal organisms and the nutritional status.

6.10.2. Hormones

Endocrinal balance is invariably altered in energy–protein undernutrition. Many hormones, including pituitary growth hormone, corticosteroids, and thyroxine, play an important role in modulating the development and magnitude of the immune response. The occurrence of high blood levels of cortisol, particularly free cortisol, in malnutrition, the susceptibility of lymphoid tissues to the lytic action of corticosteroids, and the immunodepression caused by therapeutic administration of hydrocortisone has been discussed earlier. It is likely that changes in the level and activity of thyroxine and growth hormone influence the immunocompetence of undernourished subjects.

6.10.3. Miscellaneous

The effect of nutritional deficiency on some antigen nonspecific factors of host resistance has not been studied or incompletely so (Faulk and Chandra, 1977). These include β-lysins and C-reactive protein. The important influence of concomitant infection in aggravating immunodeficiency secondary to undernutrition is discussed in Chapter 5.

7

INTERACTIONS OF NUTRITION, INFECTION, AND IMMUNE RESPONSE IN ANIMALS

7.1. GENERAL CONSIDERATIONS

7.1.1. Introduction

Practical problems in nutrition and infection are often presented by diseases in food animals. An interesting report (Crane, 1965) relates how serious dietary deficiencies predisposed a large number of beef cattle herds in California to infectious bovine rhinotracheitis and calf diarrhea. After seven years of searching, analysis of soil, feedstuffs, and animal tissues revealed deficiencies of vitamin A, phosphorus, protein, and energy. Correction of these deficiencies with food supplements and institution of appropriate immunization programs resulted in raising the weaned calf crop from 73 to 94%.

The mechanisms of such practical problems, in both food animals and in humans, can often be revealed through the use of animal models. The field of nutrition has advanced significantly in recent years because of experimental work conducted on a broad spectrum of animal species (Munro, 1971). Little research, however, has been designed to investigate the specific interactions of nutrition and disease. This has derived on the one hand from a lack of interest or expertise in infectious disease problems by nutritionists and from a lack of interest or knowledge about nutrition on the part of those studying infectious disease. Only in recent years have well-planned, controlled experimental studies shown that nutrition and infection are interrelated and that nutritional status influences the course of infectious disease in man and animals and vice versa (Scrimshaw *et al.*, 1968). However, in an extensive recent review (McFarlane, 1973) of over 200 publications on immunoglobulins, fewer than a dozen deal with animal models. This simply reflects the dearth of work and data available in animal pathology related to nutrition and infection.

7.1.2. Species Variations in Dietary Needs

In attempting to determine the effects of diet and nutritive status on the response to the stress of infection, a number of aspects must be considered. These include the species of animal, its caloric need per unit of time in relation to units of body weight and of surface area, and the normal rate of growth.

The general concept of requirements for basal metabolism was formulated during the early years of the 20th century. Basal metabolism is proportional to the surface area of an animal (Widdowson, 1970). This in turn bears directly on the animal's nutritional needs. One must use discretion, therefore, in the choice of the species used to study a specific aspect of the relationship of nutritional status to the response to infection, since the choice of species may significantly influence the results obtained. Rats or pigs, for example, grow more rapidly than man, and this fact creates some difficulty if one wishes to extrapolate results of animal experiments to the human infant when the exact dietary conditions are not stipulated. For example, consider the results shown in Table 7.1, taken from the work of McCance and Widdowson (1964). These observations indicate clearly the difference between several species of animals in the time required to double birth weights, relative to calorie and to protein intakes. The fast-growing rat and pig need fewer calories and less protein to gain a gram of body weight than the human child because they double their weight more quickly, but they need more calories

Table 7.1
Caloric and Protein Intake of Newborn to Double Weight[a]

Species	Birth weight (g)	Time to double weight (days)	Total energy intake (kcal)	Kcal intake/g gain	Total protein intake (g)	Intake/g gain (g)
Man	3,500	150	114,000	32.5	1,900	0.54
Calf	35,000	60	370,000	10.6	16,800	0.48
Pig	1,400	7	6,500	4.6	400	0.28
Rat	5	5	15	3.0	0.75	0.75

[a] From McCance and Widdowson (1964).

and protein per gram per day. This can directly affect the way the individual species responds to the stress of infection.

If an animal is not provided with proper nutrients during this period of most rapid growth, it is much more sensitive to the stress of an infectious agent. Not only does an animal require a high level of food intake during the period of rapid growth, but the foods must provide a proper balance of the various dietary ingredients. In other words, one of the most critical aspects of adequate nutrition is that of balancing the source of calories from the three major classes of ingredients. Protein is the most crucial single ingredient in the diet of rapidly growing animals and must be kept within well-defined limits if the homeostasis is to be maintained. The work of Payne (1970) shows that at weaning the growth rates per kilo of body weight differ widely between species but that, with the exception of the group of primates, growth rates are very similar when expressed in terms of metabolic body weight (Table 7.2). This means that there is a fixed relationship between protein requirements and basal energy requirements and that these must be balanced if the animal is to be in the optimum physiologic state to resist infectious disease.

Crawford (1968) demonstrated that tissue proteins of a wide variety of species are very similar in their amino acid compositions, which implies that there should be very little interspecies variation in requirements. Mitchell (1955) has suggested that some species' requirements may vary with age, however. Whereas in the young animal the amino acid content of the protoplasmic tissues would be expected to dominate the overall requirements, at maturity the needs of certain species for keratin synthesis or for synthesis of such products as wool and feathers would be of proportionately greater importance; this might then lead to a greater requirement for the sulfur-containing amino acids in the adult.

The rat is a convenient animal model for nutritional studies; its use, during the first 40 years of this century, coincided with the era of vitamins and minerals, when studies about these two classes of dietary ingredients overshadowed all the other studies.

Recent studies have revealed that two species of animals widely used in laboratory investigations (hamsters and rabbits) have nonspecific problems or diseases that surface with stress,

Table 7.2
Growth Rates of Different Species,
Weanling and Half-Adult[a]

Species	Weanling		Half-adult
	g/day/kg	g/day/kg$^{0.73}$	g/day/kg$^{0.73}$ [b]
Mouse	105	35	23
Rat	60	35	18
Guinea pig	40	27	12
Cat	30	25	10
Dog (beagle)	26	25	16
Pig	20	36	23
Rhesus monkey	5.2	5.2	1.1
Chimpanzee	6.0	8.2	1.7
Gorilla	3.7	5.4	1.3
Man	2.4	4.1	0.5

[a] From Payne (1970).
[b] Refers to metabolic body weight.

particularly stress of nutritional origin. This has little documentation in the literature, but has been observed by many people working with the animals in controlled investigations.

Two conditions peculiar to the hamster are "wet-tail," a regrettable bit of nomenclature, and the phenomenon of hibernation. Wet-tail appears to be an infectious disorder with an obscure etiology. The relationship of wet-tail to sanitation and the diet has been demonstrated (Gay, 1968). In our own experience, wet-tail and other forms of nonspecific enteritis, as well as regional ileitis or pneumonia, commonly surface when the hamster is subjected to nutritional stress, particularly vitamin A deficiency.

Rabbits are very susceptible to enteritis, pseudotuberculosis, and a respiratory condition, caused by *Pasteurella,* referred to as snuffles. Although these conditions are recognized clinically and morphologically, studies relating them to dietary imbalances or deficiencies have not been reported. On the other hand, conditions such as *Eimeria stiedie* or hepatic coccidiosis are often associated in rabbits with less than optimum nutritional status.

7.1.3. Synergism and Antagonism

Epidemiologic studies, clinical investigation, and laboratory experiments have clearly shown that malnutrition and infectious disease can be mutually aggravating; together, they can produce more serious consequences than would be expected from a summation of the independent effects of the two conditions. Such synergistic relationships exist between malnutrition and infection: (1) infections are likely to be more severe in animals with clinical or subclinical malnutrition, and (2) infectious diseases can turn borderline deficiencies into severe malnutrition.

Occasionally, the interreaction of nutrition and infection is antagonistic. That is, malnutrition may actually decrease the severity of the infectious disease of the host, but this is the exception rather than the rule. Synergism is characteristic of most diseases caused by extracellular microorganisms, while those few instances of antagonism have been associated more with intracellular agents, primarily viruses. The quantity and variety of experiments and observations recorded in the literature (Scrimshaw *et al.*, 1968) are extensive and from them the following general statements about nutrition and response of the host to infection can be made: (1) protein deficiencies generally result in synergistic effects although there are rare instances of antagonism with selected amino acid deficiencies, (2) vitamin A deficiency is regularly synergistic, (3) vitamin D deficiency often fails to show interaction but synergism has been described, (4) the complex of B vitamins seems to behave in a variable fashion depending upon the agent and the host; this complex is associated with most of the demonstrated instances of antagonism, (5) vitamin C deficiency usually is synergistic but antagonism has been recorded, (6) specific mineral deficiencies may result in either synergism or antagonism depending to some degree on the infective agents.

7.2. NUTRITION–INFECTION INTERACTIONS

7.2.1. Protein

Because protein is intimately involved in the response to stress of infection, plasma amino acid concentrations are often examined in experiments dealing with infection. Plasma amino acids have a characteristic periodicity, and some of the conflicting data in the literature result from failure to recognize this phenomenon. The metabolism of amino acids of the host is altered greatly during the course of infectious disease, but mechanisms of specific changes are not well understood. Metabolic and endocrine disturbances which occur during infection are intimately related to the protein metabolism and are highly important during critical periods of stress. An extensive list of publications based on clinical observations and rigidly controlled experimental animal studies clearly establishes the highly detrimental effects of protein malnutrition on resistance to infectious disease. It is also known that most, if not all, of the effects of vitamin deficiency on antibody formation probably result from interference with protein synthesis. For example, this relationship has been demonstrated in studies with dietary pyridoxine (Axelrod, 1958; Blackberg, 1927–28; Crane, 1965); these studies have been further supplemented by investigations in which amino acid analogues have been used. In the work reported by Madden and Whipple (1940), dogs depleted of protein by repeated plasmapheresis had greatly reduced capacity to elaborate specific antibodies, and the animals were more susceptible to infection; if they were subsequently given adequate protein, the decrease in response was reversed.

Bacterial infections have been most widely studied. DuBois and co-workers (DuBois, 1955; DuBois and Schaedler, 1958; Schaedler and DuBois, 1959) found that mice infected with *Mycobacterium tuberculosis* had a significantly higher fatality when fed 5 to 8% of casein than when fed either 8% casein plus 12% supplemental amino acids or 20% casein; the effects were synergistic (Table 7.3). Koerner *et al.* (1949), using the rat as the test animal, found that infection with *M. tuberculosis* caused the demise of all experimental animals within 150 days when they were fed diets of 15 or 25% protein, whereas 40% dietary protein resulted in no losses and very little progression of the disease. Using the same test organism, *M. tuberculosis,* in the hamster, Ratcliffe and Merrick (1957) reported that the progress of the disease was more rapid with 6% protein than with 30% protein. There was an eventual arrest and even regression of the infection when the animals were fed 30% protein. Rao and Gopalan (1958) observed that guinea pigs infected with *M. tuberculosis* had a more rapid return to positive nitrogen balance when fed 18% protein than when fed 5% protein. Wissler (1947) observed that

Table 7.3
Effect of Quality and Quantity of Protein
on Mouse Tuberculosis[a]

Treatment	Weight at infection (g)	Cumulative mortality, days postinfection			
		4	5	8	14
14 Days preinfection					
20% casein	28	1	2	3	4
8% casein	26	3	4	7	10
21% protein pellets	26	1	3	6	6
37 Days preinfection					
20% casein	28	2	3	4	6
8% casein	29	2	3	4	6
21% protein pellets	25	2	4	4	6

[a] Data extracted from DuBois (1955); DuBois and Schaedler (1958); Schaedler and DuBois (1959).

rabbits depleted of protein and infected with *Diplococcus pneumonia* had higher fatality, decreased agglutinin response, and depressed phagocytic activity. In this case, however, low caloric intake may also have influenced the response.

Siegel *et al.* (1968) confirmed the effect of dietary protein on resistance to infection. Their studies revealed that excessively high dietary protein levels, as well as low ones, adversely affected the response of chicks to infection with avian tuberculosis (Table 7.4). McGuire *et al.* (1968) demonstrated that rats fed lower than

Table 7.4
Effect of Dietary Protein on Avian Tuberculosis[a]

Treatment Dietary protein (%)	Weight (g)			Tubercles per LPF[b]		TB index
	Body	Liver	Spleen	Liver	Spleen	
6.7 Control	155	6.5	0.2	—	—	—
6.7 Infected	142	11.9	1.1	22	64	332
20.0 Control	464	14.3	1.3	—	—	—
20.0 Infected	452	24.1	4.2	9	23	314
47.5 Control	526	14.1	1.2	—	—	—
47.5 Infected	444	25.2	3.8	15	23	378

[a] From Siegel *et al.* (1968), modified.
[b] LPF = Low Power Field.

Table 7.5
Effect of Dietary Protein Levels on *Salmonella* Infection in Rats[a]

Dietary protein %	Mortality (3-day)	Urinary N excretion (mg) 24 h postinfection	Organ weight (% Body weight)		Adrenal ($\times 10^2$)	Plasma corticosterone ($\mu g/100$ ml)
			Liver	Spleen		
0 Control	—	8	2.9	0.26	3.1	38
0 Infected	15/20	27	4.4	0.56	3.8	28
12 Control	—	56	3.1	0.24	2.1	54
12 Infected	5/20	204	4.8	0.72	3.1	25
18 Control	—	164	2.9	0.25	2.0	31
18 Infected	0/20	294	4.9	0.84	4.7	25

[a] Data from McGuire *et al.* (1968).

Table 7.6
Effect of Copper and Protein Deficiencies
and RES Blockade on Survival Time of Rats
Infected with Salmonella typhimurium[a]

Treatment[b]	Average time to death (days)	Range of time to death (days)	Number of survivors (at 30 days)
Control infected	25	7–29	16
Control infected and RES[c] blockade	12	4–15	8
Copper-deficient infected	7	5–10	4
Copper-deficient infected and RES blockade	8	5–9	3
Protein-deficient infected	7	5–10	5
Protein-deficient infected and RES blockade	9	7–12	4

[a] From Newberne et al. (1968).
[b] There were 20 rats in each group at start.
[c] RES = reticuloendothelial system.

normal levels of dietary protein had less resistance to *S. typhimurium*; decreased resistance was associated with increased urinary losses of nitrogen, and derangement of the endocrine system (Table 7.5), as is almost always the case.

Our work (Newberne et al., 1968, Newberne, 1977) demonstrated that low-protein diets, and diets deficient in copper, prevented the reticuloendothelial system (RES) from responding normally to *S. typhimurium* infection in rats (Table 7.6). Clinical evidence (mortality) coincided with histopathologic observations (Figs. 7.1 through 7.4). Livers and spleens of infected animals on a control diet were more enlarged and contained many more reactive granulomas than did livers and spleens of rats given copper- or protein-deficient diets; yet the infected controls were more capable of resisting the disease than were the animals on deficient diets. A partial explanation for the response may have been provided by further experiments (Young et al., 1968) which showed that the protein metabolism of skeletal muscle of infected rats fed low-protein diets was different from that of infected controls (Tables 7.7 and 7.8); skeletal muscle is a major source of protein during stress, serving as a source of amino acids for protein synthesis as well as a source of energy. If infection depresses ribosomal protein synthesis, then perhaps less protein and amino acids are available at the critical time for use in defensive mechanisms.

There are numerous other examples of the detrimental effects of protein depletion and its relation to bacterial infections. The laboratory of Aschkenasy (Aschkenasy, 1973, 1974; Srivastava et al., 1975), studying both cell-mediated immunity and humoral

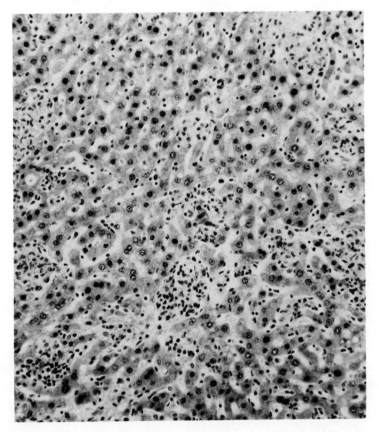

Figure 7.1. Liver from rat fed control diet and infected with *Salmonella typhimurium*. Small focal granulomas are numerous, indicating a significant response to the infection.

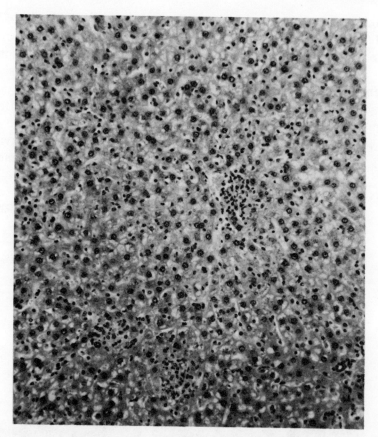

Figure 7.2. Liver from rat fed low-protein diet and infected with *Salmonella typhimurium*. The small number of granulomas indicate a failure to respond normally to the stress of infection.

immunologic mechanisms, found that protein deprivation in rats resulted in depressed levels of serum hemagglutinins. Rosette-forming capacity was unchanged. However, antibody-forming cells and mitosis of spleen cells were decreased as a result of protein deficiency; this was associated with a diminished capability for graft-versus-host reaction and was related to increased synthesis of cAMP and a prolongation of the synthesis phase of the lymphocyte cell cycle.

Passwell *et al.* (1974) used mice to demonstrate that protein deficiency resulted in slower clearance of carbon from the blood-

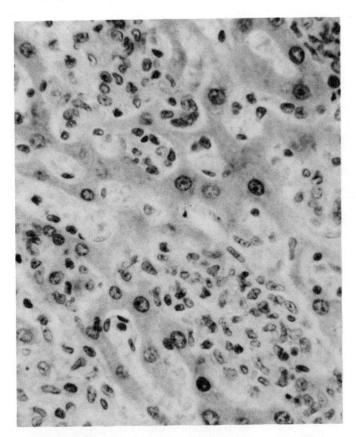

Figure 7.3. High magnification of control liver shown in Fig. 7.1. Cells of the reactive types are numerous.

Table 7.7
Effect of Dietary Protein on Muscle Nucleic Acids
and Protein Following Infection of Rats
with *Salmonella typhimurium*[a]

Parameter mg/g wet muscle	Dietary protein 18%			Dietary protein 5%		
	Days after infection			Days after infection		
	0	1	3	0	1	3
DNA	0.31	0.39	0.32	0.33	0.29	0.32
Protein	116.6	120.0	115.0	116.7	116.0	111.3
Protein/DNA	376.0	310.0	354.0	356.0	394.0	313.6

[a] From Young et al. (1968).

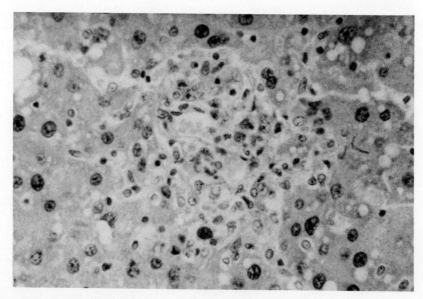

Figure 7.4. High magnification of the liver shown in Fig. 7.2. There is only a small number of reactive cells and the liver cells are vacuolated, an adverse sign.

Table 7.8
Effect of Dietary Protein on
Branched-Chain Amino Acids of Muscle
Following Infection[a]

Days postinfection	Amino acid μm/kg fresh muscle		
	Leucine	Isoleucine	Valine
5% Dietary casein			
0	100	46	100
1	35	21	36
2	29	20	35
3	125	71	20
18% Dietary casein			
0	100	67	155
1	146	100	116
2	99	68	109
3	80	51	92

[a] From Young et al. (1968).

stream, and reduced affinity of antibody to human serum transferrin, than in control mice. When mice were pair-fed isocaloric diets, low-protein diets resulted in an overall impairment of phagocytosis. Coovadia and Soothill (1976a, 1976b) employed ^{125}I-labeled polyvinyl pyrrolidone (PVP) for measurement of the clearance function of macrophages in animals deprived of protein or selected essential amino acids (Figs. 7.5 and 7.6). In Ajax mice,

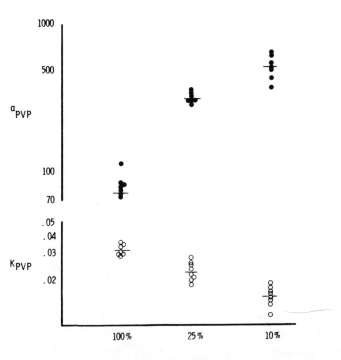

Figure 7.5. Effect of dietary tryptophan restriction on PVP clearance. Young animals of BALB/c strain were pair-fed isocaloric diets varying in content of tryptophan by selective reconstitution of the depleted diet. Changes in blood levels of ^{125}I-labeled PVP were measured over a period of 24 to 48 hours after intravenous administration. Results were expressed as the exponential rate constant K obtained from the slope of the regression line of log blood level against time and as α calculated as $K^{1/3} \times$ body weight/(liver weight + spleen weight). K_{PVP} probably estimates the overall phagocytic capacity of the animal and α_{PVP} the function of individual macrophages. From Coovadia and Soothill, 1976b.

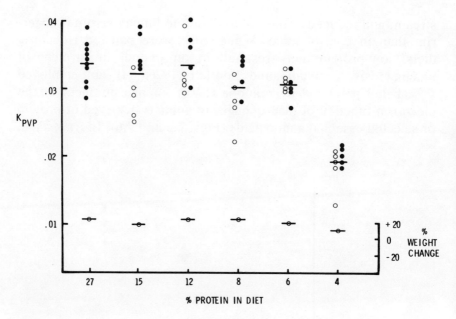

Figure 7.6. Effect of protein restriction on PVP clearance and weight in Ajax mice: ● = males, ○ = females. Experimental details as in Fig. 7.5. From Coovadia and Soothill, 1976a.

a strain with a genetically determined high clearance ability (Morgan and Soothill, 1975), there was no significant difference of PVP clearance when dietary protein ranged from 27 to 6%. At 4% casein intake, however, K_{PVP} decreased significantly. The fall and rise of K_{PVP} was observed as early as 3 days of feeding on protein-restricted or protein-replete diets respectively. The remarkable speed with which impairment and recovery of PVP clearance could be achieved with dietary manipulation suggested a rapid turnover of macrophages as a result of cell proliferation as well as the hepatic migration of phagocytic cells from another organ or tissue. Well-nourished female animals had a higher PVP clearance than males of the same strain. These sex differences disappeared when protein was restricted to 6% or less. The abolition of the sex differential of K_{PVP} on protein deprivation is not easily explained. There may be a complex relationship between the effects of protein nutrition and the known stimulatory effect of estrogens on macrophages. In malnutrition, estrogen levels are likely to be high due to reduction in hepatic degradation of the hormone which is suggested by the occurrence of gynecomastia in individuals with moderate to severe nutritional deficiency. Changes in clearance were observed without significant alteration in growth velocity. The fall in K_{PVP} suggests that protein restriction results in a reduction in the total number of phagocytosing reticuloendothelial cells. The simultaneous increase in α_{PVP} points to an increased functional capacity of residual individual macrophages. If during nutrient deprivation, macrophages and hepatocytes are lost in parallel, the remaining cells may be functionally more active, a compensatory overshoot phenomenon. If decrease in organ weight is largely a result of hepatocytic loss, there may well be a reduction of function of individual macrophages. It is also possible that calculations are fallacious if macrophage function is reduced below the threshold level of detectability of phagocytosis. Similarly, limitation of dietary tryptophan to 25% of growth-promoting standard diet or less resulted in progressive reduction of PVP clearance, and was associated with growth failure or an actual loss of weight. Liver, thymus, and spleen were significantly smaller and lighter.

These data are apparently in conflict with earlier observations. However, differences in technique of estimating reticuloendothelial cell function, the nature, size, and dose of particle challenge, genetic strain of animal, pair-feeding or *ad libitium* feeding, physical state of the diet (pellet, dry powder, or paste), total caloric intake, relative change in total body and organ weights, may all influence the modulation of macrophage clearance function by nutritional factors. Carbon has been a popular challenge particle and, in the doses commonly employed, it may both block and stimulate the reticuloendothelial system. Radiolabeled bacteria, colloidal gold, saccharated iron, colloidal thorium, or aggregated albumin, all have intrinsic fallacies. Further experience may show that ^{125}I–PVP clearance may also have some limitations.

Good and colleagues (Jose and Good, 1971, 1973; Jose *et al.*, 1973) have reported on interesting and somewhat paradoxical findings in rats and mice deprived of protein during different periods of early life. Enhancing antibody was absent in cell-mediated immunity to tumor heterografts in rats deprived of protein—a significant finding in view of the problems which enhancing antibody causes immunotherapy in cases of malignancy. These investigators have also reported a profound long-term effect of early nutritional deprivation on the developing immunological system. The more severe the early deficiency, the less likely cytotoxic cellular immunity was to reach normal levels later in life. Overall, early deprivation of protein resulted in persistent defects in cytotoxic immune function, decreased numbers of T lymphocytes in spleen, thymus, and lymph nodes, and diminished specific antibody responses. However, in another study, a range of moderate protein deficiency was found in which cellular immune responses were enhanced because of lack of blocking serum antibody. Animals on such regimens had marked inhibition of the growth of allogeneic, syngeneic, and autochthonous tumors.

Fernandes *et al.* (1976) studied the influence of low-protein diet on the immunologic function of NZB mice, an animal strain whose natural life history is characterized by progressive loss of lymphocyte proliferative response to mitogens, reduced number of T lymphocytes, splenomegaly, and Coombs-test-positive anemia. Animals fed on an isocaloric diet containing 6% casein as a source of protein showed a significant alteration in immune reactivity associated with lower weight gain and hypoproteinemia. The deprived group did not develop splenomegaly and the thymic involution was less pronounced. The striking changes in serum immunoglobulin concentrations with advancing age in NZB mice on normal protein diet were inhibited in the low-protein group. The relative number of thymus-dependent θ-positive lymphocytes in the lymph node increased initially and decreased later in life. The 6% protein diet was associated with a relative maintenance of higher response by spleen cell to T cell lectins phytohemagglutinin and concavalin A. The response to lipopolysaccharide, which stimulates B lymphocytes, declined with aging equally in the two groups. In the deprived mice, antibody production to sheep red blood cells was higher, graft-versus-host response was more vigorous, and the cytotoxicity of spleen cells was significantly greater than corresponding values in the control group. The low-protein diet delayed the development of autoimmune hemolytic anemia characterized by low hematocrit and positive

Coombs test but did not prevent it entirely. Life span was not significantly changed. The authors caution, however, that although studies with NZB mice indicated that protein deficit diminished immunocompetence, separation of genetic from nutritional influences was not remarkably clear. Good et al. (1976, 1977) have recently summarized the evidence that nutritional restriction can exert powerful influences on the immune response, the nature and severity of the effects being dependent upon the type and intensity of immunologic stimulus, and whether the nutritional deprivation is acute or chronic, moderate or severe.

The prolific laboratory of Ramalingaswami has provided an enormous amount of data relative to protein deficiency and immunocompetence (Ratnaker et al., 1972; Mathur et al., 1972; Bhuyan and Ramalingaswami, 1972; Bhuyan and Ramalingaswami, 1973; Bhuyan and Ramalingaswami, 1974). The phagocytic function of the reticuloendothelial system measured by using ^{32}P-labeled E. coli was reduced in protein-deficient rhesus monkeys. In rabbits fed a low-protein diet, staphylococcal bacteremia was rapidly fatal. The early exponential removal of bacteria was slow. The neutrophilic response was poor and transitory. Bacteria persisted and multiplied in the tissues, and focal necrotizing lesions rather than well-formed abscesses were found in various organs. Protein deficiency in guinea pigs was accompanied by marked inhibition of local and systemic immune responses to BCG immunization. The BCG nodule showed a marked delay and deficiency in the mobilization of macrophages. The draining lymph node showed little proliferation of paracortical lymphocytes, retarded transformation of epithelioid cells, absence of caseation necrosis and persistence of bacilli for a long time (Figs. 7.7 and 7.8). Tuberculin sensitivity was greatly impaired. The immune response to sheep red blood cells was depressed in protein-deficient rats and improved on injection of syngeneic X-radiated thymocytes. In subsequent studies, Krishnan et al. (1974) have observed a marked atrophy of thymus and spleen in vitamin A-deficient rats with concomitant protein–calorie malnutrition. Immune responses to sheep red blood cells were diminished as were responses to diphtheria and tetanus toxoid infections. Bhuyan et al. (1974) observed that the phagocytic and bactericidal activities in protein-deficient rabbits under the experi-

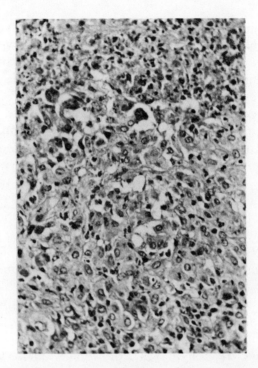

Figure 7.7. Draining lymph node of guinea pig fed high-protein diet, 21 days after BCG (H & E × 240). From Bhuyan and Ramalingaswami, 1973.

mental conditions of the study remained unimpaired. Deo et al. (1973) showed that dietary protein deficiency reduces particulate clearance by the reticuloendothelial system. Suda et al. (1976) studied the kinetics of mobilization of neutrophils and their marrow pool in protein–calorie deficiency. The pattern of appearance of ^{3}HTdR labeled leukocytes in the peripheral blood was identical in protein-deficient and control rats under basal condition. Following subcutaneous implantation of sterile glass cover slips, a higher proportion of labeled PMNs appeared earlier in the blood in the deprived animals. However, the local inflammatory exudate was poor with delayed appearance of monocytes. A reduction in the marrow pool of neutrophils was observed in protein-deficient monkeys.

Lopez et al. (1972) found that protein deficiency in young marasmic swine resulted in defects in both humoral and cellular immunity, mainly in the latter. Numerous other studies in animals have recorded important alterations in immunocompetence effected by protein deficiency, primarily in the cell-mediated arm of the system (de Pablo *et al.*, 1972; McFarlane and Hamid, 1973). In refeeding experiments, the timing of immunization in relation to nutritional supplementation is important (Price and Bell, 1976).

Malavé and Layrisse (1976) investigated the differential effect of protein restriction on immunoglobulin M and G antibody response to histocompatibility H-2 antigens upon primary and secondary stimulation with allogeneic cells in inbred mice of C57BL/6 strain. After weaning, the animals were fed *ad libitum*

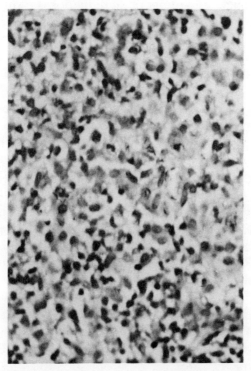

Figure 7.8. Draining lymph node of guinea pig fed low-protein diet, 21 days after BCG. There is a diffuse accumulation of macrophages and formation of immature epithelioid cells. A fair number of intracellular acid-fast bacilli are seen. Well-formed granulomas and caseation necrosis were absent (Carbolfuchsin and hematoxylin × 390). From Bhuyan and Ramalingaswami, 1973.

with a protein-deficient diet containing casein in concentration of 8%. After 6 to 7 weeks, the animals were injected intraperitoneally with 10^7 spleen cells of DBA/2 strain which differed on the H-2 locus from the recipient's cells. Employing ascitic tumor line L5178Y as target cells, the number of spleen cells forming alloantibodies was estimated in a plaque-forming cell assay system. In the protein-deprived group, there was a reduction in the total number of spleen cells and in IgG antibody-forming lymphocytes, whereas the proportion of IgM alloantibody plaque-forming cells was increased. The serum hemagglutinin titers were similar or higher during the primary response and decreased during the secondary response. Thus there was a preferential suppression of cell populations involved in the IgG response to alloantigens. An impaired feedback inhibition or a reduction in suppressor cells may explain the unchanged or higher IgM response. Since pair-feeding was not attempted and the animals fed a protein-deficient diet consumed lower amount of food than mice receiving normal diet, the effect must be attributed to protein and calorie deficiency.

A related aspect of cellular immunity is the ability to reject transplanted tissue from a genetically unrelated donor. Gautam *et al.* (1973) studied protein–calorie malnourished mice and found that their homograft survival time was significantly higher than that of controls. Jose and Good (1971) observed that protein-deficient rats underwent more rapid graft rejection and therefore seemingly exhibited enhanced cell-mediated immunity. Purkayastha *et al.* (1975) evaluated the influence of protein deficiency on rejection of skin homografts from another strain incompatible at a major locus. Protein restriction to 3% casein for 3 to 4 weeks did not alter the mean survival time of homografts. One limitation of this study was the relatively short duration of nutritional deprivation. It is known that the long-lived small lymphocytes, the key cells in thymus-dependent cell-mediated immunity, are reduced only after prolonged protein deficiency, perhaps as long as 70 days. Moreover, functionally different subpopulations of lymphocytes which play an enhancing or suppressor role in immune response may be affected to different degrees in malnutrition (Chandra, 1977a). The effect of malnutrition of the donor on the survival of skin homografts has also been evaluated. Tissues of protein-deficient rats survive longer than grafts from animals on a control diet. These data suggest that dietary restriction may modify the expression and activity of histocompatibility antigens.

Graft-versus-host reaction is a function of the presence, number, and biological activity of thymus-dependent lymphocytes, and has been used as an index of cellular immunity. Bell and Hazell (1975) injected young F_1 progeny from C57BL male × BALB/c female matings with lymphocytes of BALB/c parental strain donors, and estimated spleen weight to body weight ratio as the index of graft-versus-host reactivity of cells from both normal and experimental groups; the phenomenon was attributed to a relative increase in the number of thymus cells. The effect was not mediated by changes in corticosteroid production since adrenalectomy did not influence the response. It was suggested that nutritional deficiency in the experimental conditions of this study was associated with a decrease in the number of short-lived rapidly dividing B cells, whereas long-lived T cells were relatively unchanged.

The role of protein–calorie malnutrition as a host determinant for *Pneumocystis carinii* infection was studied by Hughes *et al.* (1974). In Sprague–Dawley rats fed a 23% protein diet, none of 15 acquired *P. carinii*, whereas 13 of 15 fed a protein-free diet died of fulminant pneumonitis caused by the organisms. Deaths from the infection were greatly increased when cortisone-treated rats were fed protein-deficient diets. Recovery of protein-deficient animals with infestation by dietary-protein replenishment demonstrated further the role of nutritional status in *P. carinii* infection and pneumonitis.

In the case of virus-induced diseases there is ample evidence that malnourished animals, particularly those deficient in protein, are less likely to resist effectively an infectious agent. Mice fed a low-protein diet and injected with infectious hepatitis agents isolated from humans had a fatality rate twice that of animals fed a higher level of protein (Ruebner and Miyai, 1961). An excellent example of virus infection in chicks is provided by Squibb (1964). The severity of Newcastle disease, measured clinically and in changes in concentrations of liver nucleic acids, was significantly increased in chicks fed either deficient or excessive amounts of protein. In our own laboratories, prenatal protein deprivation clearly affected the response of dogs to viral infection. Exposure at six months of age to distemper virus revealed a carryover to young adulthood of defective immunocompetence from prenatal deprivation of protein (Table 7.9). Post-weaning undernutrition in mice led to lymphoid atrophy, and coxsackie virus B3 infection resulted in prolonged persistence of the pathogen and increased severity of lesions (Woodruff, 1970; Woodruff and Kilbourne, 1970). Abnormally low interferon titer was found in starved animals inoculated with 10^7 plaque-forming units of the virus. However, with a larger dose of the virus (10^8 units), normal levels of interferon were produced despite marked atrophy of lymphoid organs.

Examples of interactions of malnutrition and protozoal and helminthic infestations in man and animals are widely available from the literature (Demarchi, 1958; Goldberg, 1959; Giron-Mendez, 1963; Mettrick and Munro, 1965; Platt and Heard, 1965; Ritterson and Stauber, 1949). In all cases cited, there was a marked synergistic effect between protein malnutrition and infectious disease.

7.2.2. Fat

Studies of mice infected with viable and nonviable tuberculosis vaccines showed that the addition of fatty acid mixtures to total 5% of a synthetic diet regularly increased survival time, irrespective of varying proportions of the six fatty acids (Hedgecock, 1958). Most studies, however, have emphasized the effects of excessive, rather than of deficient, dietary fat. Solotorovsky and co-workers (1961) reported that relatively severe challenging infections masked the dietary effects of fat in avian tuberculosis. With less severe infections, however, low dietary fat prolonged the median survival time of chicks and led to a reduced number of tubercles per tissue section in the spleen.

Fiser *et al.* (1972) have observed that high-fat diets fed to dogs increased considerably the severity of infection with the virus of infectious canine hepatitis. This was associated with decreased capacity of the white blood cell series to perform normally.

Table 7.9
Effect of Protein Deprivation during
Intrauterine Development of Pups on Postnatal Response
to Canine Distemper Virus

Maternal dietary treatment (g) protein/kg. body wt.		Number of litters	Average number pups per litter	Average birth weight (kg)	Average 6-month weight (kg)	Number with paralytic encephalitis
Gestation	Lactation					
1.25	3.75	8	5.0	0.261 ± 0.04	8.9 ± 0.2	26/35
3.75	3.75	6	6.5	0.307 ± 0.02	9.3 ± 0.3	12/35

Monkeys fed diets rich in polyunsaturated fatty acids developed severe diarrhea, and examination of the feces revealed high concentrations of *Entamoeba histolytica* (Scrimshaw *et al.*, 1968). It was concluded that the diet aggravated the effects of amoebic infection in monkeys.

In view of the significant effects observed in the relatively small number of studies using high-fat diets, there appears to be reason for concern for animals exposed to infectious disease when high levels of fat are included in their diets.

7.2.3. Carbohydrates

The relationship of carbohydrates to immunocompetence has received little attention, even though the calories provided by carbohydrates are undoubtedly essential to good immune response. There is only one report in the literature (Chandler *et al.*, 1950); it deals with *Hymenolepis diminuta* infection in rats. Sucrose, corn starch, agar, or glucose was used as the source of carbohydrate; the only observation of significance was that the parasites were smallest in animals receiving sucrose and largest in those consuming corn starch. This might be seen as correlating with the observation in rats that sucrose alone results in less growth and diminished clinical health than does a mixture of carbohydrates. Obviously, there is need for research in this area of nutrition-infection interactions and carbohydrates.

7.2.4. Vitamins

It now seems clear that most of the specific vitamin deficiencies do interfere with antibody production in experimental animals. The early report of Blackberg (1927-28) was one of the first experimental investigations with adequate controls. In this investigation, killed typhoid bacilli or small doses of live bacilli injected into rats deficient in vitamins A and D and the B complex resulted in measurably lower titers of agglutinin and bacteriolysin than those observed in control animals.

7.2.4.1. Vitamin A

Vitamin A concentrations in the livers of people living in different parts of the United States (Raica et al., 1972), as well as in the world at large, vary widely; serum levels of vitamin A were low in about 25% of the population examined in the U.S. national nutrition survey.

The appearance of nyctalopia (night blindness), decreased vision in partial darkness resulting from inadequate levels of vitamin A, is, unless corrected, the forerunner of more serious consequences. This early symptom is usually overlooked, and vitamin A deficiency is suspected only when derangement of epithelial tissue has become evident. Xerophthalmia, respiratory diseases of chickens and other animals, urolithiasis, and other urogenital disorders, result primarily from alterations in epithelial structures which provide a portal of entry for infectious agents. Numerous publications describe how experimental animals, deficient in vitamin A, frequently develop spontaneous infections. Probably it was a result of such observations that vitamin A came to be known as the "anti-infectious" vitamin. Early work on vitamin A deficiency (Guggenheim and Buechler, 1947) established beyond doubt the interrelationship of vitamin A and the integrity of the epithelial tissues. Squamous metaplasia of the epithelial lining in the parotid gland, followed by similar changes in other surface epithelia, was established as characteristic of vitamin A deficiency. Presumably, interference with the integrity of the epithelium, particularly secretory epithelium, interferes with the ability of the animal to resist infection and allows the infectious agent to gain entrance to the tissues. Although the classical lesions and symptoms of vitamin A deficiency are rarely seen, it is imperative that the possible presence of subclinical deficiency be kept in mind when the ability of an animal to respond to superimposed infectious disease is being evaluated.

Over 50 published investigations deal with diseases of bacterial, viral, or protozoal origin in which vitamin A deficiency resulted in greater frequency, severity, or mortality. When the prevalence, tissue change, or mortality has been used as a criterion, tuberculosis has been shown to be more severe in a variety of hosts when there is a concurrent vitamin A deficiency. Although many species have been studied (Solotorovsky et al., 1961; Getz et al., 1951; Sriramachari and Gopalan, 1958; Finkelstein, 1931-32), controls were not always provided.

Guggenheim and Buechler (1947) reported synergism between *Salmonella* infection and vitamin A deficiency in mice and rats with a lowered resistance to infection being encountered before clinical signs of the deficiency appeared. These investigators concluded that decreased food intake and probably protein deficiency were also involved.

Sherman and Burtis (1927-28), Bradford (1928), and Turner et al. (1930) found that vitamin A deficiency in rats greatly increased the susceptibility to infection; this susceptibility continued long after the rats were returned to supplemented diets.

The work of Wissler (1947), Crane (1965), Newberne et al. (1968), and of Bang et al. (1972) further supports an important role for vitamin A in the immune response. Chicks deprived of vitamin A for varying periods of time after hatching had depleted lymphocyte and plasma cell populations in the upper respiratory tract; this coincided with injury to epithelium and failure to replace damaged cells. These modifications, along with depletion of lymphocytes of the bursa of Fabricius, predisposed the vitamin A-deficient chicks to infection with Newcastle disease virus. More than 100 times the concentration of Newcastle disease virus was recovered from throat swabs from the vitamin A-deficient group than was recovered from controls, and cell-mediated immune response was depressed. These observations clearly indicate a morphologic and functional suppression of lymphocytes in vitamin A-deficient chicks, probably resulting from a deranged interplay of epithelial cells and lymphocytes.

Viral infections (Weaver, 1946; Webb, 1916; Gratzl et al., 1963; Panda et al., 1964; Roos et al., 1946) are more severe in vitamin A-deficient animals. Yaeger and Miller (1963) have demonstrated that rats conditioned for 13 weeks on a vitamin A-deficient diet are significantly more susceptible to *Trypanosoma cruzi* infections and to bacterial infections than are pair-fed controls.

Vitamin A-deficient dogs exposed to infestation with *Toxocara canis* and *Toxascaris leonina* had five times as many parasites at autopsy as did dogs fed an adequate diet and exposed at the same time (Wright, 1935). Roundworms in pigs (*Ascaris lumbricodes*) seemed initially to do more damage in animals that were vitamin A deficient; eventually, however, the signs of infestation equalized in experimental and control animals (Claphan, 1934). Hiraishi (1927) found that he could infect young pigs with the human strain of roundworm only when the pigs were deficient in vitamin A. Observations of 3 vitamin A-deficient and 2 well-fed cats showed that naturally acquired *Diplidium canium* was far more intense in animals fed the deficient diet (MacKay, 1921). Rats infected with *Schistosoma mansoni* destroyed all of the parasites in the liver and lungs if they were given adequate vitamin A-supplemented diets, whereas those deficient in vitamin A failed completely to destroy the organisms (Krakower *et al.*, 1940).

Thus, it seems that vitamin A deficiency is consistently synergistic with infectious disease. The strong suspicion that marginal or deficient vitamin A may well be a problem in animals around the world requires further investigation.

Although there are a number of references in the literature to vitamin A deficiency precipitated from marginal deficiencies by superimposed infection in humans (Scrimshaw *et al.*, 1968), there are no references to well-designed studies in animals. We have observed this phenomenon in the course of other studies, however (Newberne *et al.*, 1968). Male weanling rats (15 per group) were fed a diet that contained either 0.3 µg retinyl acetate per gram of diet or 3.0 µg per gram of diet. After 1 month, half of each group were infected with *S. typhimurium,* and three days after infection they were sampled for serum and liver concentrations of vitamin A. Table 7.10 lists results of the determinations which clearly indicate that the infection sharply depleted the stores of vitamin A. To further corroborate the effect, rats that survived (3 or 4 per group) were held for an additional 48 days, sacrificed, and examined for evidence of histopathologic alterations. Figures 7.9 and 7.10 illustrate that the infected, marginal vitamin A group had evidence of vitamin A deficiency while their infected, supplemented cohorts did not.

Beisel has further indicated that malnutrition can occur as a consequence of stress in animals and man and points out the

Table 7.10
Effects of Infection on Status of Vitamin A
in Rats, Three Days Postinfection

Dietary treatment	Infected with *Salmonella typhimurium*	Serum (μg/100 ml)[b]	Liver (μmg/g)[b]	Mortality (72 hours)
Control diet[a] (3.0 μg/g diet)	0	49 ± 7	96 ± 10	0
Retinyl acetate[a]	+	27 ± 5	61 ± 12	3/15
Marginal diet[a] (0.3 μg/g diet)	0	23 ± 4	11 ± 3	0
Retinyl acetate[a]	+	2 ± 1	1.7 ± 1	7/15

[a] Fifteen rats per group.
[b] Mean ± SD on 5 animals only.

needs and preventative measures which should be considered under such circumstances (Beisel, 1977).

7.2.4.2. Other Fat-Soluble Vitamins

Little literature deals with the effects of vitamins D, E, and K on resistance to infectious disease. Synergism was observed in two studies of experimental salmonellosis in rachitic rats. In both cases, the reduced resistance of vitamin-deficient animals was restored by adding vitamin D to the diet (Robertson and Ross, 1932; McClung and Winter, 1932). Young pigs that developed rickets when fed barley diets deficient in vitamin D readily acquired a fatal salmonellosis (Manninger, 1928), whereas well-nourished animals showed no evidence of clinical disease.

A deficiency of vitamin E in the hamster and rat appears to permit the acid-fast bacilli of human leprosy to grow in both species; under conditions of adequate dietary supplementation the bacilli are completely refractive. This finding indicates that inadequate vitamin E might be a complicating factor in some infectious disease (Mason and Bergel, 1955). Another report indicates that deficiency of vitamin E prevented the normal resistance of weanling mice to intranasal innoculation with the virus of vesicular stomatitis (Sabin and Duffy, 1940). Thus, deficiencies of vitamins D and E appear to be synergistic with infectious disease under some conditions, but there is a need for studies on the interac-

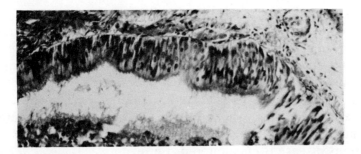

Figure 7.9. Primary bronchus from rat fed control diet with adequate vitamin A and infected with *Salmonella typhimurium*. Note tall columnar cells lining the passage.

tions of these vitamins and on interactions of vitamin K with infectious agents.

7.2.4.3. Water-Soluble Vitamins

For a complete description of the many reports in the literature relative to water-soluble vitamins and infectious disease, the reader is referred to the extensive review by Scrimshaw *et al.* (1968). Much of the work with the B complex was done before all of the individual B vitamins were isolated and identified; more recent investigations, however, have dealt with individual vitamins of the B complex. Well-designed, controlled experiments have shown that insufficiencies of ascorbic acid,

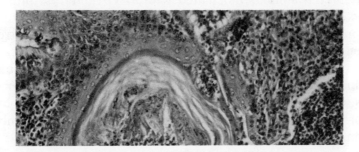

Figure 7.10. Primary bronchus from rat fed marginal level of vitamin A and infected with *Salmonella typhimurium*. The infection precipitated a deficiency of vitamin A evidenced by squamous metaplasia, hyperkeratosis, and bronchopneumonia.

thiamin, riboflavin, niacin, pyridoxine, pantothenic acid, folic acid, vitamin B_{12}, inositol, choline, and paraminobenzoic acid all produce, in general, a synergistic action with infectious disease in man and animals (Scrimshaw *et al.*, 1968). Under practical conditions one rarely encounters a deficiency of a single water-soluble vitamin such as the studies referred to above; clearly, complex deficiencies, whether acute and severe or only marginal, can have a more highly detrimental effect on the hosts when infectious disease is encountered. For example, Bang *et al.* (1973) have demonstrated the effect of vitamin B deficiency in Newcastle disease in lymphoid cells in chickens. Some more recent data on water-soluble vitamins and the immune response are of sufficient significance to describe in some detail.

7.2.4.3a. Vitamin B_6. Early experiments on albino rats deficient in vitamin B_6 showed impairment of antibody response to human erythrocytes, diphtheria toxoid (Pruzansky and Axelrod, 1955) and influenza virus (Axelrod and Hopper, 1960). Secondary responses were inhibited to a greater extent than primary ones (Axelrod, 1958), and nutritional supplementation during later stages of the experiment did not improve antibody production. When the antagonist deoxypyridoxine was administered to rats shortly before the second immunizing dose of diphtheria toxoid, the resulting induced pyridoxine deficiency interfered with the secondary antibody response even after a normal primary response had occurred. Similar observations were made on vitamin B_6-deficient animals treated with deoxypyridoxine. In nutritionally deprived animals inoculated with *Mycobacterium tuberculosis* or BCG vaccine, delayed cutaneous hypersensitivity to purified protein derivative was depressed, even though the correlates of sensitization were demonstrable *in vitro* (Trakatellis *et al.*, 1963). In the studies by Blackberg (1927–28) mentioned earlier, pyridoxine deficiency decreased the response to sheep erythrocytes, and hemagglutinin response in rats inoculated with human erythrocytes was impaired by dietary deficiency of a number of B-complex vitamins. Rejection of skin homo-transplants across the major histocompatibility barrier was delayed in deficient animals (Axelrod *et al.*, 1958), and specific tolerance to donor antigens and consequent acceptance of skin grafts could be achieved by injection of the donor's spleen cells into vitamin B_6-deficient mice.

Robson and Schwarz (1975a), studying the effect of vitamin B_6 deficiency on cellular immunity in rats by employing the mixed lymphocyte reaction (MLR), found that B_6 deficiency resulted in a reduction in the relative and absolute number of lymphocytes in the peripheral blood and thoracic lymph. Lymphocytes from the thoracic duct were collected through a surgical cervical fistula. Unidirectional MLR was evaluated with stimulating cells from F_1 hybrids of Lewis and Brown Norway rat strains (which differ at the major Ag–B histocompatibility locus) and by measurement of incorporated ^{3}H-thymidine. The uptake *in vitro* of ^{3}H-uridine by small lymphocytes from the deprived group was low, possibly reflecting a change in lymphocyte subpopulations as well as impaired function. Mononuclear cells from vitamin B_6-deficient rats had a reduced ability to respond *in vitro* to antigenically foreign lymphocytes, and the size of cutaneous reaction following lymphoid cell transfer into abdominal wall, an *in vivo* correlate of MLR, was correspondingly reduced.

Recent studies demonstrated that deficiencies of calories (Chandra, 1975g) in rats impair immune responses in the first and second generation progeny. Vitamin B_6 deficiency induced in pregnant rats reduces the size of the thymus in the dams as well as their fetuses (Davis *et al.*, 1970). Robson and Schwarz (1975b) showed that vitamin B_6 deficiency *in utero* was associated with higher mortality in rat litters exposed to a bacterial infection; function of thoracic duct lymphocytes was reduced, as judged by MLR and graft-versus-host reaction following cell transfer into the skin.

The mechanisms involved in the immunosuppressive effect of vitamin B_6 deficiency are not clear. Some evidence, however, suggests that the lack of vitamin B_6 impairs nucleic acid synthesis, with consequent deleterious influence on protein synthesis, cell division, and repair. The adverse effect of vitamin B_6 deficiency during pregnancy on immunocompetence in offspring may result from derangement or reduction in numbers of the cells that are progenitors of thymus-dependent lymphocytes. Much remains to be done in this important area of nutrition and immunocompetence.

7.2.4.3b. Folic Acid. Studies in our laboratory (Williams *et al.*, 1975) demonstrated that rats deprived of folic acid postweaning had deranged immunocompetence and decreased resistance to infection with *S. typhimurium*. Wistar–Lewis rats were fed either a control or a folate-free diet from weaning. After 4 weeks, they were sensitized with skin grafts from Brown Norway (BN) rats. Following initial sensitization, all animals received weekly i.p. injections of 3×10^7 BN thymocytes. Compared with controls at age 3 months, the folate-deficient rats had mild megaloblastic changes in the bone marrow and small intestine, a mild decrease in cellularity in both the thymus and the thymic-dependent areas of the spleen, decreased serum folate levels, and decreased overall body weight. Hematocrit levels and weights of spleen and thymus did not differ significantly from those of controls. However, the cytotoxic activity *in vitro* of splenic lymphocytes from folate-deficient animals exposed to BN thymocytes decreased significantly (Table 7.11), as did sensitivity to stimulation by the T cell mitogen, phytohemagglutinin (PHA) (Table 7.12). The number of T cells in the spleen and peripheral blood of folate-deficient rats was significantly lower than in controls; the number of T cells in the thymus of deficient rats was also lower than in the control group, but not significantly so; labeling of T cells by ^{3}H-uridine was decreased.

The difference in mortality between these experimental and control rats infected under identical conditions, together with the inability of lymphocytes from folate-deficient animals to respond normally *in vitro* during tests for T cell competence, point to a defect in the immune system, most probably in the cell-mediated response. The decreased response of lymphocytes to PHA can be explained by several possible mechanisms: (1) inability of the T lymphocytes to undergo transformation to a blast form, (2) inability to synthesize DNA and to reproduce after undergoing

Table 7.11
Cytotoxicity Assay of Splenic Lymphocytes
from Folate-Deficient Rats[a]

	Control	Folate deficient	Significance
Cytotoxicity (% killing)	29.1 ± 3.7 (16.3 to 40.0) $n = 7$	5.2 ± 1.5 (0 to 13.9) $n = 8$	$p < 0.005$

[a] From Williams *et al.* (1975).

Table 7.12
^{3}H-Thymidine Uptake by PHA-Stimulated Spleen Cells[a]

	Control (cpm)	Folate deficient (cpm)	Significance
PHA stimulation			
No PHA	1315 ± 312	1405 ± 102	
	(887 to 1701)	(1789 to 8248)	
+ PHA	19,398 ± 1014	4263 ± 579	
		(1789 to 8248)	
Stimulation index	14.8	3.0	$p > 0.005$
	$n = 8$	$n = 10$	

[a] From Williams et al. (1975).

blast transformation, or (3) overall reduction in the number of T lymphocytes. Likewise, the decreased cytotoxic capacity of lymphocytes from folate-deficient rats may involve the afferent and/or effector stages of the cell-mediated response. An afferent defect, specifically the inability of an animal's lymphocytes to recognize or process antigen, could prevent the animal from becoming sensitized to allogeneic antigen, despite repeated exposures. Alternatively, the defect may involve the effector arm of the cell-mediated immune response, in that folate-deficient lymphocytes, though sensitized, may be unable to generate adequate numbers of cytotoxic cells to respond effectively to the allogeneic target cells.

7.2.4.3c. Vitamin B_{12} Studies. We have fed rats diets containing a deficit, a normal recommended amount, or an excess of vitamin B_{12} during pregnancy and lactation, then placed their progeny on a diet containing the normal recommended level of vitamin B_{12} for up to 1 year (Newberne and Gebhardt, 1973).

Offspring exposed to abnormally high amounts of the vitamin during the pre- and perinatal periods had higher birth weights than controls and were more resistant to *S. typhimurium* infection (Table 7.13). In a recent series of experiments repeating the above studies, an examination of spleen and peripheral lymphocytes revealed that rats exposed to unusually high amounts of vitamin B_{12} had an increased response to PHA stimulation. These observations indicate that vitamin B_{12} alone may be highly important in

Table 7.13
Mean Values for Litter Size, Birth Weight, Body Weights,
Vitamin B_{12} Activity, and Mortality of Progeny from Rats
Fed Low, High, or Control Levels of Vitamin B_{12}
during Pregnancy and Lactation[a]

Parameter	Deficient	Control	Excess
Number of litters	26	44	57
Litter size[b]	7.9 ± 2.0	9.7 ± 1.5	9.9 ± 1.2
Birth weight[b]	4.7 ± 0.3	5.8 ± 0.2	6.2 ± 0.3
Weaning weight[b]	34 ± 4	41 ± 4	51 ± 3
1-Year weight, males[b]	398 ± 14	463 ± 9	510 ± 9
B_{12} concentration (ng/g)			
Dam	64 ± 12	132 ± 32	320 ± 70
Newborn	8 ± 4	25 ± 17	74 ± 28
Positive *Salmonella* cultures from blood			
18-hour postinfection	17/20	16/20	18/20
48-hour postinfection	16/20	11/20	5/20
10-Day cumulative mortality	33/40	21/40	9/40

[a] From Newberne and Gebhardt (1973).
[b] Mean ± SD.

development of immunocompetence and that needs may increase significantly during the period in which the thymolymphatic system is developing rapidly, perhaps also in gestation.

We have conducted studies, similar to those described above, in mice of the C57BL/6 strain which were fed either a vitamin B_{12}-deficient or -supplemented diet. The striking reduction in numbers of antibody-forming cells and diminished transformation of lymphocytes by PHA or lipopolysaccharide (LPS) (Table 7.14) parallel the results observed in studies with vitamin B_{12} deficiency alone, which can affect the immune reaction of thymus-derived lymphocytes in both rats and mice, and certainly influences the response to infectious disease.

7.2.4.3d. Choline–Methionine. We became interested in dietary lipotropes because marginal, subclinical deficits are now recognized in women of Western populations formerly considered to be well fed: pregnant women, alcoholics, those consuming oral contraceptives or anticonvulsants, some food faddists, the poor. We induced subclinical deficiencies in animal models by lowering dietary concentrations of four lipotropic factors (choline, methionine, vitamin B_{12}, folic acid), together or singly, and examining

Table 7.14
Responses of Lymphocytes from Normal and
Vitamin B_{12}-Deficient Mice to Sheep Red Blood Cells
and Phytohemagglutinin (PHA)[a]

Treatment	Antibody-forming cells/10^6 spleen cells	Serum antibody titer hemagglutinin	−PHA[b]	+PHA[b]	LPS[b]
Control diet	21.4 ± 3.0	1 : 1	3,181 ± 460	68,312 ± 6,130	52,160 ± 2,734
B_{12}-deficient	6.0 ± 1.2	1 : 1	1,684 ± 209	43,120 ± 4,103	19,680 ± 2,463

[a] 1×10^6 cells from each animal in 0.5 ml medium were incubated with or without mitogen for 48 h; ^{3}H-thymidine was added, and the experiment terminated at 72 h. From Gebhardt and Newberne (1974).
[b] Counts per min per 10^6 cells.

the effects on defense mechanisms, particularly cell-mediated immunity. (Newberne *et al.*, 1970a; Newberne *et al.*, 1970b; Newberne and Wilson, 1972; Chanarin *et al.*, 1968).

We have also studied rats that were littered to dams that had been given only marginal levels of choline and methionine (Gebhardt and Newberne, 1974). The thymus glands of the prenatally malnourished pups were smaller than the thymuses of pups adequately nourished *in utero*. Furthermore, we observed decreased cellular density in the thymuses, and often in the peripheral lymphoid systems (spleen and lymph nodes) (Figs. 7.11 and 7.12) of the experimental rats. The experimental rats appeared normal at birth and grew almost as well as the controls; yet when, as young adults, they were infected with *S. typhimurium,* the 5-day postinfection mortality rate was higher in the marginally deprived group than in the control group (3/20 versus 14/20). Total white cell counts and serum proteins were lower in the deprived

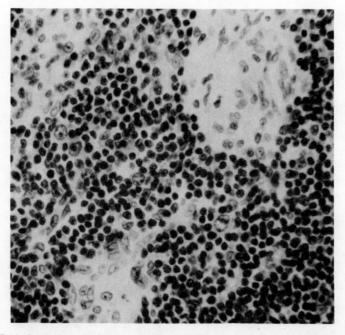

Figure 7.11. Lymph node from normal, adequately nourished rat. Large numbers of lymphocytes are present.

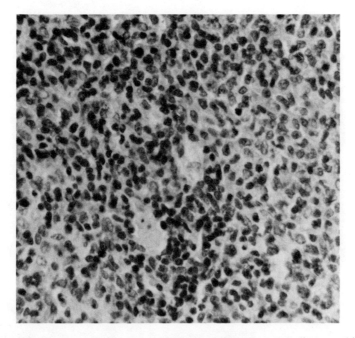

Figure 7.12. Lymph node from rat deprived of lipotropes (choline, methionine) in perinatal period. Note diminished number of lymphocytes.

rats. Thus, maternal lipotrope deficiency had weakened offspring to infection, without clinical evidence of a defect until the challenge by infection.

Following observations that marginal deprivation of lipotropes *in utero* resulted in hypoplasia of the thymolymphatic system (Figs. 7.13 and 7.14), we examined the capacity of such deprived animals to be immunized against sheep red blood cells (SRBC) and of their lymphocytes to respond to mitogen stimulation. Clearly, the offspring of dams fed the marginal lipotrope diet did not respond as well as normal, age-matched rats to SRBC, and serum antibody titers were lower in the deprived progeny (Tables 7.15 and 7.16). This paralleled clinical response to infection, shown in Table 7.17.

Spleen cells from deprived rats responded poorly to stimulation with the T cell mitogen, phytohemagglutinin (PHA); but the thymus cells of deprived animals responded as well as cells from

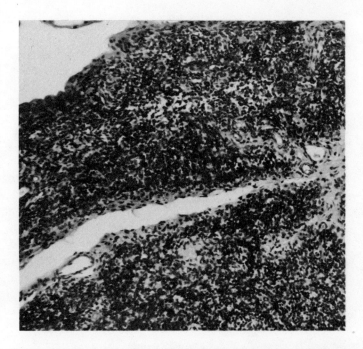

Figure 7.13. Thymus gland from rat fed normal diet and infected with *Salmonella typhimurium*. Although the lobes are slightly diminished in size, there is a significant cellular density.

Table 7.15
Phytohemagglutinin (PHA) Stimulation of Spleen and Thymus Cells from Normal and Lipotrope-Deprived Progeny[a]

	^{3}H-Thymidine incorporation (cpm)	
	−PHA	+PHA
Normal progeny spleen	5,686	65,380
	(1,171–9,819)	(49,500–74,700)
Deprived progeny spleen	3,348	6,154
	(1,683–5,120)	(4,200–8,340)
Normal progeny thymus	4,756	29,420
	(1,806–5,700)	(21,200–34,504)
Deprived progeny thymus	4,025	31,563
	(2,740–7,000)	(24,320–38,701)

[a] 1×10^6 cells from each animal in 0.5 ml medium were incubated with or without mitogen for 48 h; ^{3}H-thymidine was added, and the experiment terminated at 72 h. From Gebhardt and Newberne (1974).

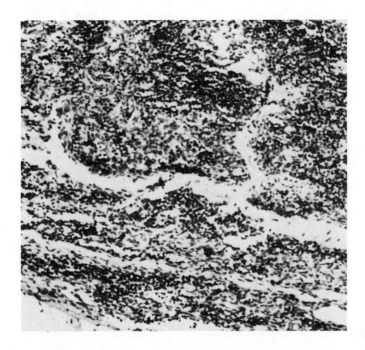

Figure 7.14. Thymus gland from rat deprived of lipotropes and infected with *Salmonella typhimurium*. Note severe loss of thymocytes.

Table 7.16
Anti-SRBC Responses of Rats Born
to Lipotrope-Deprived Mothers[a]

Cell and serum source	Antibody-forming cells/10^6 spleen cells[b]	Serum antibody level[c]	
		Hemolytic titer	Hemagglutinating titer
Normal offspring	275 ± 13	1 : 2,048	1 : 256
Marginal lipotrope offspring	37 ± 8	1 : 32	1 : 8

[a] From Gebhardt and Newberne (1974).
[b] The values given represent the mean and standard error of counts obtained from four separate assay plates. The total number of spleen cells per animal was not found to be significantly different in any of the experiments. For this experiment the cell numbers were: normal, 400 × 10^6; marginal lipotrope offspring, 392 × 10^6.
[c] The values given were obtained using serum samples from individual rats of each type. Five other experiments involving two animals of each type (normal and marginal lipotrope) per experiment yielded similar data.

Table 7.17
Effect of Lipotropes on Birth Weight, Growth, and Response
to Infection with *Salmonella typhimurium* in Rats[a]

Lipotrope status		Number of animals	Average weight at infection (g)[b]	Cumulative mortality (%) Days postinfection		
Gestation and lactation	Postweaning			7	14	30
Deficient	Deficient	17	233 ± 6	100	100	100
Deficient	Complete	20	240 ± 8	80	100	100
Marginal	Marginal	62	248 ± 5	50	80	91
Marginal	Complete	65	285 ± 7	40	73	90
Moderate	Moderate	70	260 ± 3	47	64	71
Moderate	Complete	60	308 ± 4	20	20	35
Complete	Complete	60	303 ± 4	10	20	25

[a] From Newberne (1975).
[b] Mean ± SE.

control rats did. The reasons for this might relate to failure of the thymus to process the peripheral lymphocytes properly (Gebhardt and Newberne, 1974). Retarded maturation and/or migration might explain the relatively poor antibody response of lipotrope-deprived animals to SRBC, a probably T-dependent antigen in rats (Borum, 1972; Newberne, 1975).

The results obtained with deprived progeny are also compatible with a quantitative deficit of PHA-responsive cells, although it is possible that the cells are present in normal numbers but cannot respond to PHA because of some internal defect. Only in cases of severe maternal malnutrition are gross and histological deficiencies apparent in the tissues of the offspring (Figs. 7.12 and 7.14).

Inhibition of a fully competent and mature T cell system in a malnourished, diseased organism may bear no direct parallel to the T cell deficiency seen in the progeny of malnourished mothers. A greater similarity is likely to exist between the mechanism responsible for immune deficiency in the offspring of deprived rats and the mechanism responsible for cell-mediated immune deficiency and thymic dysplasia in undernourished children (Smythe *et al.*, 1971).

Further study of possible B cell deficiencies in the offspring of undernourished mothers is required. Offspring of nutritionally deprived mothers have fewer antibody-producing cells following immunization with SRBC. Assuming this antigen is T-dependent in the dose used, the apparent deficiency of B cells may actually reflect a helper T cell deficit. Gebhardt and Newberne (1974) found that the progeny of lipotrope-deprived mothers and normal rats have equivalent numbers of lipopolysaccharide-responsive cells in their spleens, and that this cell population is not greatly expanded in the normal or experimental groups by oral infection with *S. typhimurium*, the source of the lipopolysaccharide. Additional studies may pinpoint the nature of T cell deficits in offspring of nutritionally deprived mothers.

The derangement of the thymolymphatic system resulting from pre- and perinatal deprivation of lipotropes is probably mediated through defective development of the thymus during intrauterine and early postnatal life, in part, and later by failure of normal DNA synthesis by lymphocytes and other components of the reticuloendothelial system. In pregnant animals, deficits of

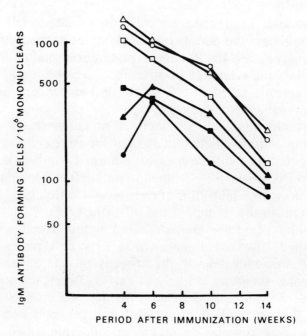

Figure 7.15. IgM antibody-forming cells in the spleen of starved (closed symbols) and control (open symbols) rats and in their progeny. Three-week-old rats were subjected to partial starvation for 6 weeks. One batch from each experimental and control group (F_0 generation, ○ ●) was immunized and the animals were killed after 4, 6, 10, or 14 days. Another batch of female rats from the starved and control groups were mated with healthy male animals. On weaning, the litter (F_1 generation, △ ▲) was given free access to food for 6 weeks. Some of these animals were immunized with SRBC and studied. Others were mated with healthy males and the progeny (F_2 generation, □ ■) were evaluated at age 4 weeks. Means are shown, based on data from Chandra, 1975g. Copyright 1975, American Association for Advancement of Science.

either vitamin B_{12} or folic acid result in diminished cell-mediated immunity; it therefore appears that methyl group metabolism is intimately involved in development of an adequate body defense mechanism.

7.2.4.3e. Vitamin C. Cameron and Pauling (1974) have presented a review of studies of the effect of vitamin C upon the immune response and remarked that occasionally a deleterious effect of vitamin C deficiency on cell-mediated immunity may be observed. Kalden and Guthy (1972) found that prescorbutic guinea pigs have markedly reduced immunocompetence and easily tolerate allografts, which appear to be caused by lymphocyte depletion. Species variations in obligate requirements of vitamin C may be critical in such experiments.

7.2.5. Calories

It is recognized that calorie or energy deficiency is the most prevalent form of human malnutrition in the world. If animal models are to be used to provide answers to questions arising from clinical problems, the investigation of the effects of calorie deficit on immune responsiveness are most pertinent. We have documented that an overall reduction of food intake in mice and rats results in thymolymphatic atrophy, reduced cell-mediated immunity, and impaired antibody response to sheep red blood cells, a T cell-dependent antigen (Chandra *et al.*, 1973). Interestingly, the immune response is depressed in the first and second generation offspring of deprived dams (Figs. 7.15 and 7.16)

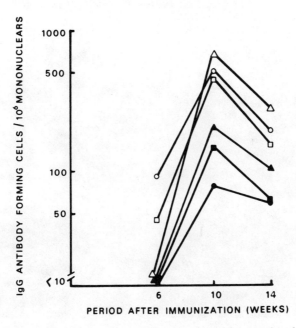

Figure 7.16. IgG antibody-forming cells in the spleen of starved and control rats, and their progeny. Experimental details and symbols as in Fig. 7.15.

(Chandra, 1975g). These observations are relevant to the reduction in cell-mediated immunity found in low-birth-weight, small-for-gestation infants (Chandra, 1974c, 1975c, 1976f, 1977c; Ferguson *et al.*, 1974; Chandra *et al.*, 1977c).

7.2.6. Minerals

Mineral deficiencies are likely to be encountered in several animal species, and are most often seen in domestic animals that are used for food production. Iron deficiency has been a problem in swine for decades, but in recent years parenteral injections have been used to offset the iron deficiency that is prevalent in newborn pigs. Recently it has been demonstrated that mineral deficiency in human populations of the United States may be more common than was originally suspected. For this reason, it is important to understand more about the interaction of mineral deficiency and infectious disease in animal models.

Iron nutrition exerts a direct effect on the morphology and function of lymphoid organs, including the thymus. Rats deprived

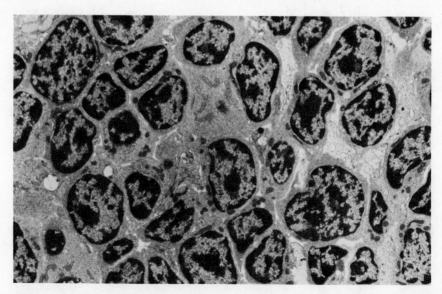

Figure 7.17. Electronmicrograph of the normal rat thymus. × 6384. Reduced 44% for reproduction. From Chandra *et al.*, 1977d.

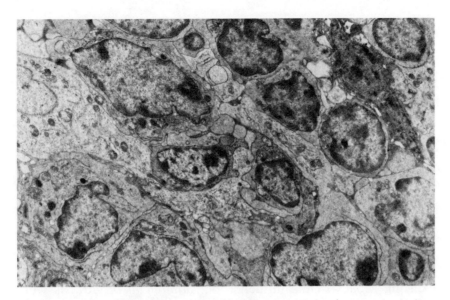

Figure 7.18. Electron micrograph of the thymus of an animal depleted of iron by administration of desferoxomine mesylate 10 mg/100 g body weight for 7 days. × 9576. Reduced 44% for reproduction. From Chandra et al., 1977d.

of iron show a mild atrophy of the lymphoid tissues with prominent depletion of lymphocytes (Figs. 7.17 and 7.18) (Chandra et al., 1977d). The number of splenic antibody-forming cell responses to immunization with sheep erythrocytes is reduced. This is compatible with human data on immunocompetence in iron deficiency (Chandra, 1976a).

Cort and Otto (1940) and Otto (1965) have reviewed in detail the evidence from animal studies for an interaction between iron deficiency and parasitic disease. Foster (1936) found that iron deficiency greatly increased the disability of dogs and cats infested with internal parasites. He induced iron deficiency by feeding a whole-milk diet grossly deficient in iron, or by repeatedly bleeding the animal. Infection was induced by feeding the parasite *Ancylostoma canium* to the animals twice a week.

Since most animals consume diets that contain at least some iron, the occurrence of acute iron deficiency is unlikely. However, an insidious, prolonged marginal depletion can also affect response to infection. In a study of 603 infants fed a proprietory baby formula containing vitamins and 12 mg of iron per quart, the incidence of respiratory infections was approximately half that of a group of 445 infants fed an evaporated milk formula with supplemental vitamins but no additional iron (Andelman and Sered, 1966).

Recent data, derived from a rat model, indicate the importance and the complexity of iron status and supplementation in relation to infectious disease. In initial trials, Baggs and Miller (1973, 1974) found that rats fed iron-deficient diets after weaning were more susceptible to challenge with *S. typhimurium*, with the greatest morbidity and mortality observed in marginally rather than severely deficient animals. Preweaning iron deficiency decreased the rats' ability to resist the stress of infection, even if a period of nutritional rehabilitation intervened. Phagocytes isolated from the peritoneal cavity of iron-deficient rats were as capable of exerting a bactericidal influence on *Salmonella in vitro* as were cells isolated from iron-adequate animals. Furthermore, myeloperoxidase (MPO)-containing cells isolated from iron-deficient rats could not be distinguished from those of iron-adequate animals with regard to the amount of MPO per cell or bacterial capacity (Table 7.18). However, the iron-deficient rats had fewer MPO-containing cells in the lamina propria and submucosa. In fact, they had less MPO, regardless of how the data were expressed (total MPO in the gut, MPO per gram protein, or MPO per kilogram body weight). In other words, rats appear unable to produce MPO-containing cells in sufficient quantity to withstand the stress of infection. The authors propose a variety of explanations for the type of response observed. Mature female rats subjected to an acute, severe lack of dietary iron gave birth to rats with impaired resistance. A group of animals weaned from these deficient dams to a diet containing 35 ppm iron had hematocrits statistically indistinguishable from the controls', but were more susceptible to infection. This may represent a latent state of iron deficiency, undetectable by clinical means, but evident when the animals are stressed.

The observation that several deficient animals receiving no dietary iron (0 ppm) were relatively resistant to challenge with *Salmonella* is puzzling. Weinberg's hypothesis (1971), that the

Table 7.18
In Vitro Characteristics of MPO-Containing Cells from Iron-Adequate and Iron-Deficient Rats[a]

	Hematocrit	MPO/cell	Extracellular log of reduction in number of Salmonella	Intracellular log of reduction in number of Salmonella
		MGU/10^6 cells[b]		
Iron adequate →	40	2.00 ± 0.55[c] (5)[d]	0.34 ± 0.34 (14)	0.83 ± 0.51 (14)
Iron deficient→	30	3.77 ± 2.08 (8)	0.59 ± 0.13 (3)	1.25 ± 0.57 (3)

[a] From Baggs and Miller (1974).
[b] MGU = milliguaiacol unit.
[c] Mean ± SD.
[d] Figures in parentheses indicate number of animals.

amount of iron in the intestinal lumen determines the outcome of oral infection, was tested by growing *Salmonella* in defined media *in vitro*. No impairment in growth rate was noted when the media contained 0.5 ppm iron (Table 7.18). However, Wawszkiewicz *et al.* (1971) have reported a "*Salmonella* resistance factor," or pacifarin, produced by enteric bacteria. This material is synthesized in iron-restricted media, and protects mice against parenteral *Salmonella typhimurium*.

We believe that oral infection of the rat with a virulent enteropathogen provides a realistic and useful model for pre- and perinatal effects of nutrition and infection. To the extent that such experimental results are indicative of processes occurring in man, the finding that marginally deficient rats are least able to resist *Salmonella* challenge is especially significant. This observation is at variance with the intuitive bases of many nutrition programs, that any improvement in nutritional status is laudable. Based on these experiments, slight improvement is not sufficient; nutritional adequacy is the only acceptable goal. There is a critical need for both clinical and basic nutrition studies into the potential problem of iron deficiency and infection.

Results obtained by us (Newberne *et al.*, 1968) seem to imply that diets deficient in copper prevent the reticuloendothelial system from responding normally to infectious disease. Table 7.6 illustrates the effect of deficient dietary copper and protein on the response of the rat to infection with *S. typhimurium*. As mentioned earlier, copper- and protein-deficient rats were less capable of resisting disease than were infected controls, even though the livers (Figs. 7.1 and 7.3) and spleens of the controls contained more reactive granulomas (Newberne et al., 1968).

There is very little published about the role of calcium and phosphorus in the interactions of nutrition and infection except where it is mentioned in literature on vitamin D and rickets. Gaafar and Ackert (1958) reported that chicks deficient in calcium and phosphorus were much more susceptible to the infestation with *Ascaridia galli* and that they had larger and more numerous parasites under these conditions. Lichstein et al. (1946) used mouse encephalomyelitis virus to show that infection was not affected by deficiencies of calcium and phosphorus; Foster and associates (Foster et al., 1949) found similar responses in mice infected with polio virus Type 2. In view of the very limited number of studies and observations, it is impossible to evaluate effects of these two highly important structural minerals on the resistance to infection.

Potassium deficiency (Woods et al., 1961) resulted in a marked increase in frequency of pyelonephritis and kidney abcesses in rats exposed to infection with *Escherichia coli*. Since the kidney is the prime target for morphologic and metabolic changes associated with potassium deficiency, this is not unexpected. However, very little is known about the influence of marginal potassium deficiencies which occur without notable clinical evidence.

Other minerals, including manganese, selenium, and cobalt, have been briefly examined, but few findings of any consequence have come out of work in this area. Therefore, no general statement can be made about the effect of mineral deficiencies on resistance to infectious disease. It is becoming increasingly apparent, however, that deficiencies of potassium and other minor electrolytes may be of considerable consequence to the response to a broad spectrum of diseases afflicting man and animals.

7.2.7. Effects of Excessive Nutrition

7.2.7.1. Distemper in Dogs

Overnutrition and obesity in animals and response to infectious disease has come under serious scrutiny in research laboratories only recently. Numerous clinicians have observed that extremely well-nourished or obese animals respond very acutely to infectious disease and appear to have a more serious problem than those that are more normally fed or even those that are slightly underfed. We (Newberne, 1966) used dogs in studies of obesity, since like man dogs tend to overeat and become obese if given the opportunity and if restricted in physical activities. Responses to the virus of canine distemper were studied. Canine distemper virus was chosen because it is the most important viral disease of the dog and well-defined strains of the virus are available for experimental use. Furthermore, similarities exist between distemper virus and at least one viral disorder in man (measles). In this series of studies, litter-mate beagles about 6 months old were housed in individual metabolism cages and fed a balanced dog ration formulated in our own kitchens. Only the daily amount of diet offered to each group was different; the high-caloric groups received 100 kcal/kg body weight per day, the low-caloric group 50 kcal/kg body weight per day, and the normal or intermediate group 75 kcal/kg body weight per day. We spent considerable time determining the exact levels of intake for maintenance, slight weight loss, and obesity over a 6-week period. Under the conditions outlined above, the overfed animals increased their food intake for about 5 weeks and then reached a plateau with much slower increments in body weight.

Over a period of about 5 years, and with many different groups of experimental animals, it was found that the group fed 50 kcal/kg body weight per day would lose about 1 kg of body weight over the 6-week conditioning period. Those fed the 75 kcal/kg of body weight would just maintain their body weight over the 6-week period, and the group fed 100 kcal/kg of body weight would gain an average of 35 to 65% of their original body weight, and could be classified as obese. At the end of 6 weeks, the dogs were innoculated with a strain of the Snyder–Hill distemper virus.

Table 7.19
Mortality, Body Weight, and Survival Time of
Distemper-Infected Dogs Fed Varying Levels
of Food 6 Weeks Preinfection[a]

Food	Average body weight (kg)		Average survival time (days)	Incidence of paralytic encephalitis	
	Initial	At infection		Number	%
Control diet, infected 75 kcal/kg/day	9.2	9.2	10	11/15	74
Low calorie diet, infected 50 kcal/kg/day	9.3	8.1	14	5/16	31
High calorie diet, infected 100 kcal/kg/day	9.6	13.0	8	20/23	87

[a] From Newberne (1966).

During the early studies with canine distemper virus, the time at which the animal developed paralytic encephalitis was chosen as the end point. The results (Table 7.19), repeated several times, clearly point out the synergism of obesity and the infectious agent; obesity was clearly a form of malnutrition as devastating as that usually associated with sharply decreased food intake.

The animals fed the low caloric intake responded very well clinically; however, none of the three groups could be classified according to the way they responded with antibody titers to the infectious agent. There was no correlation between serum-neutralizing antibody and paralytic encephalitis. For this reason, we feel that serologic tests in use today to determine the status of distemper infection are at best only partially reliable. In our studies, animals with exceedingly high antibody titers often developed paralysis.

Total plasma protein tended to decrease in all groups following infection. This partially reflected dehydration in the high-intake group but not in the low-intake group. The overfed animals were incapable of producing the level of serum γ-globulins that was observed in the control and the low-fed groups. For this reason, it might be hypothesized that overnutrition has some depressing effect on the reticuloendothelial system and its capacity to produce protective immunoglobulins. In this case, with distemper infection, it should be pointed out that the virus grows extremely well in the lymphoid tissue and this correlated well with the depression and destruction of lymphatic elements by the virus. During this series of studies, it became evident that the most serious disturbances in the biologic system related to protein metabolism, which in turn is related to the conditioning effects of diet prior to exposure.

A most striking observation in all infected groups was the loss of nitrogen. Total urinary nitrogen losses after about 7 days of infection approximated 35 g. This enormous amount came largely from the muscle. Obese dogs appeared to lose more protein nitrogen in a shorter time than did the control or the low-fed groups. Moreover the nitrogen losses were in part essential liver amino acids. High-fed, obese dogs lost a higher proportion of essential amino acids than did low-fed animals, the latter tending to conserve this class of amino acids. These losses resulted in important changes in the ratios of essential to nonessential free amino acids.

Since most of the protein lost during the period soon after exposure to infection comes from the muscle, a study of the changes in nucleic acid and protein in muscle was conducted. Both ribonucleic acid and protein were found to decrease within 24 hours of infection in both the high- and the low-fed animals but stayed down in the obese group. Later studies indicated that it took a few days before the levels returned to normal. We concluded that the obese animal was less capable of managing its protein metabolism during the stress of infection than normally fed infected animals.

7.2.7.2. Salmonella in Dogs

More recent investigations in dogs using *Salmonella* infection (Fiser *et al.*, 1972; Newberne and Williams, 1970) have confirmed that obese animals or animals fed high levels of dietary fat have decreased capacity to resist bacterial infection and exhibit more serious clinical disease. Williams *et al.* (1972) have demonstrated a clear positive correlation between decreased total free amino acids in serum and a favorable clinical response to the infectious disease. Interestingly, the change in free amino acid concentrations precedes, by 12 to 24 hours, clinical evidence of disease. An increase in the serum's total free amino acids following exposure to *Salmonella* infection was equated with a poor subsequent clinical response.

Beisel (1977) and the group at Fort Dietrick have isolated a factor from leukocytes from infected animals which results in duplication of many of the clinical and biochemical observations reported in obese, infected animals.

7.2.8. Vitamin A

The influence of excessive intake of vitamins on certain *in vivo* functions dependent on cell-mediated immunity has been studied, Jurin and Tannock (1972) noted that injection of vitamin A for the 5 days preceding or following grafting, or starting on the 6th day after grafting, significantly reduced the mean rejection time of skin grafts in mice. A similar study with guinea pigs by Uhr *et al.* (1963) showed that the expression of several types of immunological and nonspecific inflammatory skin reactions was suppressed by acute hypervitaminosis A.

7.3. SUMMARY

The obscure nature of the interactions of nutrition and infection is slowly being resolved. Knowledge of this area will aid early detection of infection and therapeutic reversal of the course of disease during its incipient stages. Clearly, prenatal as well as postnatal malnutrition result in deranged metabolic and clinical responses to infectious agents; this may help explain the wide variation in response to infection seen in groups of animals with no observable physical differences.

8

BIOLOGICAL IMPLICATIONS

8.1. INFECTION-RELATED MORBIDITY AND MORTALITY

The extensive clinical and epidemiologic data attesting to the frequency and severity of infections in the undernourished have been reviewed in Chapter 1 (Scrimshaw et al., 1968; Chandra, 1976b, 1978a). Some studies suggest that the pattern of infections may be different in malnourished and well-nourished groups, in that gram-negative septicemia, tuberculosis, herpes simplex, and candidiasis occur frequently in the former group. Depression of cell-mediated immunity, secretory antibody response, and polymorphonuclear leukocyte function may be the significant pathogenetic determinants involved (Chapter 6). Primary immunodeficiency due to defective development of the thymus or phagocyte dysfunction is characterized by a similar pattern of infections (Soothill, 1975).

For many viral and enterobacterial diseases, recovery and immunity to infection on subsequent exposure depend largely on antibody response in external secretions. Such response is reduced in malnourished subjects (Chandra, 1975f; Sirisinha et al., 1975) permitting continued replication and shedding of the pathogen, thereby prolonging the period of contagiousness. Reduced mucosal immunity may also permit systemic spread; this may underlie the frequent occurrence of gram-negative bacteremia in malnutrition.

A dichotomy between serum and secretory response may be followed by severe Arthus phenomenon-like reactions or immune complex formation when such an individual is exposed to the pathogen again, as has been seen in persons immunized with inactivated measles vaccine. Antigenemia in the presence of high titers of serum antibody induces immune complex formation. Such a mechanism might underlie fulminant measles infection in malnourished subjects who may have had a subclinical infection earlier.

Anorexia nervosa represents a special form of chronic undernutrition occasionally seen in clinical practice. The total amount of energy intake is markedly reduced. Infections are not unusually frequent but paronychia and staphylococcal skin infections may often be seen (Dally, 1969). Occasionally, severe infections cause death in patients with anorexia nervosa (Warren and Wiele, 1973).

8.2. POSTOPERATIVE SEPSIS

The determinants of infection during postoperative periods are ill understood. In hospitalized patients, depressed cell-mediated immunity causally associated with semistarvation (Bistrian et al., 1975) may influence the development and progression of sepsis. The extent of tissue trauma, duration of the surgical procedure, amount of blood lost, type of anesthesia employed, and nature of the underlying disease can affect the immune response in the immediate post-operative period (Chandra, 1977i). This is associated with transiently negative nitrogen balance and hormonal changes, both of which modulate immune reactivity. Recent observations point to surprisingly frequent occurrence of malnutrition in surgical patients (Chandra, 1977i; Hill et al., 1977). Antigen nonspecific factors of host resistance, for example phagocyte function and mucociliary transport, are also depressed by surgery. The data on the effect of surgery and anesthesia on immune response has been presented and reviewed (Jubert et al., 1973; Duncan and Cullen, 1976).

8.3. PARENTERAL HYPERALIMENTATION AND INFECTION

In individuals whose nutrition cannot be sustained by dietary intake and gastrointestinal digestion and absorption, intravenous hyperalimentation is often resorted to. There are numerous problems of administration through this route of all the nutrients essential for the human body. It seems possible that inadvertent deficiencies of some nutrients can occur frequently. The frequency of bacterial or mycotic septicemia in patients undergoing parenteral hyperalimentation varies from 8 to 50% (Parkinson *et al.*, 1972). Propensity to infection has been related to the duration of intravenous cathetarization and infusion (Curry and Quie, 1971). There is some evidence to suggest that nutritional imbalance associated with intravenous hyperalimentation may impair immune response and thereby increase susceptibility to infection (Chandra, 1977b). Several other factors, including hyperosmolar environment detrimental to leukocyte function, presence of an indwelling catheter acting as foreign body, contamination of fluids, are also important in the genesis of infection in such critically ill patients. Craddock *et al.* (1974) investigated the effect of severe hypophosphatemia, a frequent complication in cachectic patients infused with dextrose and amino acids, on phagocyte function. Experimentally induced severe hypophosphatemia resulted in depression of chemotactic, phagocytic, and bactericidal activity of granulocytes. There was a simultaneous reduction in leukocyte ATP concentration. The chemotactic abnormality was corrected by phosphate supplementation *in vivo* or by incubation of leukocytes with adenosine and phosphate *in vitro*, both in animals and in a hyperalimented patient. This study provided an additional example of nutritional control of granulocyte function. It is likely that other nutrients, common and uncommon, may modulate immune responsiveness to an extent which is clinically pertinent.

8.4. INTERGENERATIONAL EFFECTS OF UNDERNUTRITION, IMPAIRED IMMUNOCOMPETENCE, AND INFECTION

Nutritional deficiency during fetal life or early postnatal period can impair physical growth. The developmental retardation may reduce the total growth potential of the individual. Infections during pregnancy also affect fetal and postnatal growth. The additive effects of nutritional deprivation and prenatal infection are devastating for the infant. And a handicapped infant grows up to be a short-statured adult. Maternal short stature is recognized to be an important determinant of intrauterine growth retardation. Thus a vicious cycle is set up.

Involution of the thymus seen in the fetuses of undernourished socioeconomically poor mothers (Naeye *et al.*, 1971, 1973) and the demonstration of impaired postnatal immunocompetence in small-for-gestation infants (Chandra, 1974c, 1975c, 1977f; Ferguson *et al.*, 1974) confirms that the immunity function of the infant may be seriously compromised by prenatal environmental factors. Recent data suggest that decreased cell-mediated immune response of small-for-gestation infants may persist for several months (Chandra *et al.*, 1977c). Primary or secondary immunodeficiency in early life has serious repercussions, both short-term and long-term, on morbidity and mortality.

In experimental animals, deficiency of calories, protein, lipotrope, iron, or pyridoxine during gestation has been shown to impair immune response of the offspring (Baggs and Miller, 1973; Gebhardt and Newberne, 1974; Chandra 1975g; Robson and Schwarz, 1975b). In one study, nutritional deprivation of rats resulted in reduction in the number and function of antibody-forming cells in the spleen of first and second generation offspring (Chandra, 1975g).

The mechanisms of the effect of nutritional deprivation on the immune response of the offspring are not clearly understood. Malnutrition during gestation reduces the size, lymphocyte number, DNA, and protein synthesis in the lymphoid organs. This may be the result of decreased availability of essential nutrients important for the synthesis of proteins and for cell proliferation. Additionally, stress hormonal factors and metabolic products may be transferred across the placenta and influence cell number and function in lymphoid tissues. In the critical first few days after birth, the amount of milk produced by the experimentally deprived dam may be reduced. An additional factor may be a suboptimal quality of milk. Information on this critical determinant of nutrition in early life is lacking. Litter size alters milk production, the total yield rising in proportion to number of litter mates from 2–12. It is possible, however, that the amount of milk potentially available has no relationship to the amount that each pup elects to consume, the latter being regulated by the duration and frequency of suckling, stomach capacity, physical activity, and metabolic activity. The highly artificial laboratory environment in which dams are studied during experiments is far from optimal and may not even be reproducible.

The human relevance of experimental data in animals must necessarily be interpreted with caution. However, studies in man show that fetal growth retardation results in lymphoid involution and impaired immunocompetence, and that these abnormalities may be more severe and longer lasting than if nutritional deficiency occurs after birth.

8.5. NUTRITIONAL DEFICIENCY, IMMUNOPATHOLOGIC DISEASE, AND AGING

The loss of subcutaneous fat, anemia, and cachexia characteristic of a majority of the elderly points to the possible occurrence of nutritional deficiency in individuals at the end of their life span. This period in age is also characterized by frequent infections, autoimmune diseases, neoplasia, and infantile disease patterns. These features suggest an impaired immunocompetence which has been confirmed in several studies. Both cell-mediated and humoral immune functions decline gradually with advancing age (Walford, 1969; Makinodan et al., 1971; Sigel and Good, 1972). In the premature senescence syndrome of progeria, there is a marked impairment of the functions of T lymphocytes and neutrophils (Fig. 8.1). There is little doubt that the three age-related events are intimately if not causally related.

At all ages, there appears to be a correlation between immune deficiency and the propensity to produce autoantibodies and malignancy. The immunodeficiency may be primary or, in the case of laboratory animals, experimentally induced by thymectomy. Recent theories suggest that aging results from a decline in immune homeostatic mechanisms (Walford, 1970). Burnet (1970) proposed the concept that a progressive breakdown of the individual's ability to determine self-recognition leads to the destructive processes of autoimmunity and cancer, both of which contribute to death in old age. There is increasing evidence to suggest that

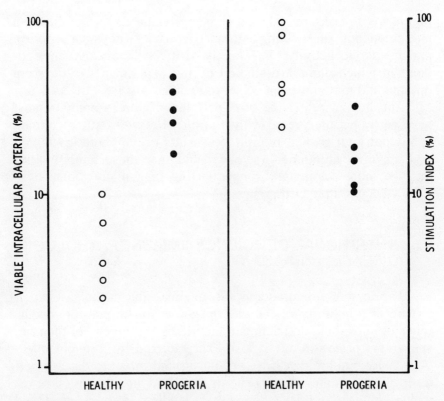

Figure 8.1. Lymphocyte proliferation induced by PHA and bacterial killing capacity of neutrophils in progeria and age-sex matched healthy controls.

the development of autoimmunity and malignancy is related to the major histocompatibility complex and immune response (Ir) genes. Burch (1968) presented a unified approach to the fundamental problems of growth, disease, and aging.

Recent epidemiologic surveys have discovered an iceberg of nutritional deficiencies in senior citizens (Nutrition Canada, 1973). The adverse effects of undernutrition, gross or subclinical, on the immune system have been documented (Chapter 6). In old age, there is a reduced incidence of delayed hypersensitivity reactions to microbial and mycotic antigens, low numbers of circulating T cells, and impaired proliferation of lymphocytes in response to mitogens and alloantigens. Some aspects of phagocyte function are also impaired. An increase in the ratio of B/T lymphocytes might heighten vulnerability to immunologic insults normally handled by cellular immunity (Greenberg and Yunis, 1972). It is possible that the slow progressive decline in immunity function associated with senescence may well be due to the presence of gross or subtle deficiencies of one or more nutrients. Malnutrition alters the hormonal milieu and the inhibitory effect of the latter on immune response is recognized. There is suggestive evidence that the immune imbalance of aging can be altered by endocrinal and dietary manipulations. These observations and the postulated interactions of nutrition, immunity, aging, autoimmunity, and cancer are fundamental to the basic questions of longevity and survival.

8.6. CANCER

The complex relationships between diet, nutrition, immunity, and cancer are receiving increased attention. Several epidemiological observations suggest that dietary nutritional factors are among the important environmental influences which may determine the recognized changes in cancer incidence that occur with migration of individuals and populations. It is important to bear in mind that since there is no single etiopathogenetic factor which underlies all cancers, it is quite likely that different types of neoplasm may be susceptible to dietary manipulation to a varying extent. In experimental animals, diet and tumor growth are intimately linked (Clayson, 1975).

Early studies stemmed from the hypothesis that tumor cells divide and grow by deriving nutrients from the host whose nutritional status may possibly affect tumorogenesis. It has been shown that chronic calorie restriction reduces the incidence and delays the appearance of many but not all tumors (Tannenbaum and Silverstone, 1957; Ross and Bras, 1971; Shils, 1973; Homburger, 1974). Calorie undernutrition inhibits mitotic activity. In the case of mammary cancer in mice, caloric deprivation may influence neoplastic proliferation by reducing the production of estrogen, a cocarcinogen. Changes in levels of adrenocorticoid and other hormones are likely to alter the genesis and growth of tumors. On the other hand, obesity and tumor risk often run parallel (Tannenbaum, 1959; Ross and Bras, 1971). Increased total amount of fat and a high proportion of polyunsaturated fats in the diet enhance the susceptibility to cancer (Wynder and Reddy, 1975). However, the effect of increased fat ingestion on cancer frequency is difficult to dissect out from the effect of total caloric intake. It is postulated that fat may act as a solvent for ingested carcinogen thereby altering the rate of carcinogen transfer (Rose *et al.*, 1974). The fatty acid fraction may exert an independent effect on the developing cancer cell. A diet high in polyunsaturated fatty acids promotes synthesis of bile salts which provide the substrate for carcinogen-forming microorganisms implicated in the etiology of colonic cancer (Pearce and Dayton, 1971). Other studies have failed to confirm such an association (Ederer *et al.*, 1971).

Differences in the protein content of diet may increase or decrease the incidence of different tumors (Ross and Bras, 1973) and operate through altered hormonal and immunological factors. Animal diets deficient in protein and amino acid may enhance tumor-inhibitory cellular immunity by reducing "blocking" humoral factors (Jose and Good, 1973). Excess of retinoids (Maugh, 1974) and deficiency of riboflavin appear to have anticarcinogenic activity, whereas deficiencies of lipotropes (folic acid, vitamin B_{12}, choline, methionine) and of iodine and magnesium may enhance the growth of certain tumors. Cameron and Pauling (1973) suggested the provocative hypothesis, largely untested to date, that ascorbic acid increases hyaluronidase which suppresses neoplastic proliferation and invasiveness.

Burkitt's (1971) widely publicized hypothesis that a low-fiber diet predisposes to colonic cancer by causing fecal stasis, the increased transit time allowing longer contact with dietary carcinogens, bacterial growth, and conversion of bile salts into potential pathogens, has some epidemiologic support. However, the pathogenetic factors and the very definition of fiber are subjects of current debate. Other dietary constituents may also alter intestinal microflora, bile salt metabolism, and metabolic enzymic function of gut epithelium, and contribute to tumor susceptibility.

Primary lymphomas of the small intestine in developing and developed countries exhibit unique differences (Dutz *et al.*, 1971).

This cancer in developing countries occurs two to three decades earlier in life than in developed areas of the world (Banihashemi et al., 1973). The preferential site of its occurrence within the gut is also different. Lymphoma may be associated and etiologically linked with heavy alpha chain disease (Rambaud and Matuchansky, 1973). It has been hypothesized (Dutz, 1975) that repeated gastrointestinal infections result in continued villous atrophy and lymphoid hyperplasia with IgA production. Chronic damage in an undernourished individual with impaired immunity is postulated to lead to mutations, secretion of heavy alpha chain, and malignancy. Alcantara and Speckmann (1976) have summarized the available data which suggest that the modulating influence of dietary intake and nutritional status may take place through specific effects on gut bacteria and substrates for bacterial metabolism, microsomal mixed-function oxidase system, hormonal balance, immunocompetence, nutrient needs for cell multiplication, and the extent and duration of exposure to the carcinogen(s). It is possible that morphological (Ramalingaswami, 1964) and immunological (Chandra, 1975e, 1975f) changes in the gut of malnourished individuals may permit an increased absorption of carcinogens. Cunningham (1976) found a correlation between animal protein consumption and lymphoma mortality, and postulated that enhanced absorption of antigens through the gastrointestinal mucous membrane results in chronic stimulation of lymphoid tissue and malignancy. This concept is supported by experimental evidence in laboratory animals and the increased incidence of cancer in celiac disease in which antigen absorption and antibody synthesis are exaggerated. Genetic susceptibility and other environmental factors such as contamination of food with oncogenic viruses may further increase cancer risk.

The etiopathogenetic role of dietary factors and nutrition in the causation of cancer has been extensively reviewed in several major conferences.*

---

* Wendell H. Griffith Memorial Symposium on Nutrition and Cancer, April 15, 1975, Atlantic City, New Jersey (published in *Fed. Proc., Fed. Am. Soc. Exp. Biol.*, May, 1976); Conference on Nutrition in the Causation of Cancer, May 19-22, 1975, Key Biscayne, Florida (published in *Cancer Res.*, November, 1975).

In patients with malignancies, another facet of nutrition–immunity–infection interaction is also seen. The incidence of infection may be related to the individual's immunocompetence, which in turn depends in part on nutritional status (Chapter 6). Bistrian *et al.* (1975) detected impaired cell-mediated immunity in hospitalized patients, which was attributed to semistarvation. Hughes *et al.* (1974) found a high incidence of *Pneumocystis carinii* infection in patients with lymphoreticular malignancy; the presence of infection correlated with a lower level of serum albumin. Our unpublished data (Chandra, 1977i) suggests that the propensity of leukemic children to develop infections is related to the status of nutrition and immunocompetence of the patient.

8.7. AUTOIMMUNITY AND ALLERGY

Immunodeficiency, especially that of the secretory immune system, is associated with a high incidence of autoimmunity and atopy. It has been suggested that impaired exclusion of antigens at the mucosal level may lead to chronic and marked stimulation of systemic lymphoid tissue, such as the forbidden clones directed against the self-antigens and IgE-producing cells. It is not yet known whether children who are malnourished early in life, including those of low birth weight, who have a poor secretory IgA antibody production, are more susceptible to developing autoimmune and allergic disorders. The numerous variables which would have a significant bearing on the issue make the design of such a study extremely difficult. Preliminary data (Chandra, 1977i) suggest that undernourished children have a higher incidence of autoallergic antibodies in their sera. Smythe *et al.* (1971) found Coombs's direct antiglobulin test to be frequently positive in malnourished children and attributed it to the presence of C4, other complement components, and immunoglobulin on the surface of red cells. The clinical significance of these findings is not clear.

8.8. FOOD ANTIBODIES

Antibodies to common food antigens have been detected in the serum of healthy infants and of individuals with a variety of gastrointestinal and other systemic disorders. The common pathogenetic denominator is a failure of prevention of mucosal penetration by antigens. This may be the consequence of structural, digestive functional, and immunological changes in the gut, which permit absorption of dietary proteins, and of reduced phagocytic function of reticuloendothelial system. In energy–protein undernutrition the mucous membrane of the intestine is thin with a marked villous atrophy, pancreatic and intestinal digestive functions are reduced, and the secretory antibody response is impaired (Chandra, 1975f, 1976b, 1977c; Sirisinha *et al.*, 1975). All these factors are likely to facilitate freer absorption of large-molecular-weight proteins which are not trapped by impaired Kupfer cell function, and would then have an access to the systemic lymphoid tissues, which form food antibodies.

Antibodies to multiple dietary proteins have been detected in the serum of undernourished children (Chandra, 1975e, 1977f). The incidence increased when new constituents of food were introduced into the diet. Antibody activity was found mainly in the IgG and IgA classes. On ingestion of food items to which antibodies were demonstrated, no untoward symptom occurred nor was complement activation observed *in vivo*.

Several pathogenetic mechanisms may contribute to the frequent occurrence of food antibodies in malnourished children. Nutritional deficiency produces gross and histomorphologic alterations of intestinal mucosa, the villous height is markedly reduced, the epithelial cells are cuboidal and atypical. The mucosal changes vary in severity and may be indistinguishable from typical coeliac histology and are reversible on nutritional rehabilitation. The thin atrophic gut may thus be more permeable to large-molecular-weight proteins which could be absorbed intact without prior digestion. The frequent occurrence of antibodies to milk proteins in cow's-milk-fed infants is a reflection of the physiological phenomenon of increased permeability of the intestine in the first few weeks of life. Infants who are initially breast fed have a much reduced antibody response when they ultimately receive bovine milk, compared to infants given cow's milk from birth, which indicates that the lymphoid cells of the former had been primed. In infants who are small-for-gestational-age, indicating fetal malnutrition, there is a high frequency of food antibodies (Chandra, 1974c, 1975c, 1977f).

Malnutrition impairs digestive processes in the small intestine. There is atrophy of the pancreas with corresponding reduction in the output of tryptic and lipolytic enzymes. The flat intestinal mucous membrane has low levels of disaccharidases and possibly of other enzymes. Impaired digestion permits the continued presence of potentially antigenic protein molecules which may then be absorbed intact through the thin mucosa.

Secretory antibodies are important for exclusion of antigens at the mucosal level. In undernutrition, the gut-associated lymphoid structures including the tonsils are shrunken and the secretory antibody response to viral vaccines is reduced (Chandra, 1975f). Impaired local immunity in malnourished individuals may permit the passage of antigens more freely in those with a normal mucosal immune system.

Nutritional deficiency is associated with impaired function of polymorphonuclear leukocytes and of macrophages. The reduced "scavenger" function of the phagocytic system may also play a role in development of antibodies to antigens absorbed from the gut. In patients with liver disease, in whom portal systemic shunts permit bypass of the hepatic reticuloendothelial system, there is a high incidence of antibodies to enterobacterial antigens and to dietary food antigens (Triger and Wright, 1973).

The immunopathogenetic significance of food antibodies is debatable. The absence of specific IgE antibodies to milk proteins, lack of complement activation on ingestion of milk, and the absence of untoward symptoms on ingestion of food items to which antibodies were demonstrated, point against any allergic attributes of these antibodies.

8.9. PROPHYLACTIC IMMUNIZATION

Since undernutrition impairs the immune response, the efficacy of prophylactic immunization could be seriously jeopardized in the individual and in the community affected by nutritional deficiency. Following BCG vaccination, tuberculin conversion occurs less frequently in malnourished subjects compared with healthy controls (Chapter 6). This is not due to reduced cutaneous response at the time of evaluation, but depends upon the nutritional status of the vaccinee. There are conflicting reports on the rate of seroconversion and serum antibody titer after administration of oral polio vaccine. If an adequate quantity and number of doses have been given and there is no interference by other enteroviruses, the serum antibody response is probably adequate in the majority of children suffering from malnutrition. This is also true of measles, tetanus, and diphtheria vaccines. The agglutinin titer after primary immunization with killed *Salmonella typhi* is, however, lower in undernourished children. The only study which evaluated antibody response to yellow fever vaccine in a small number of children with kwashiorkor showed a reduced antibody level.

Recent data indicate that the production of secretory IgA antibody in response to live attenuated polio virus and measles vaccines is impaired significantly in malnutrition (Chandra, 1975f). Since immunity to both these viral infections is dependent largely on the local mucosal barrier, the protection achieved by such vaccines may be seriously limited in populations with rampant malnutrition. Further, seroconversion without an adequate secretory antibody response may predispose to the development of adverse reactions on subsequent exposure to the live virus.

Does a lower titer of antibody response in nutritional deficiency imply reduced reinfection immunity and inadequate protection from subsequent challenge? The answer to this crucial question must await prospective epidemiologic studies. Nevertheless, the variable data on immune response in undernutrition are not in themselves a sufficient reason to deny or unnecessarily delay much-needed and useful immunization. Vaccine-induced active immunity is one of the principal mechanisms of resistance to infectious diseases. In the light of the above-mentioned observations, however, it may be wise to give nutritional supplements at the same time as immunization in undernourished subjects, or to plan vaccination campaigns in the light of local knowledge regarding the ages, seasons, and other factors when the nutritional state of children is at its best.

8.10. IMMUNOPOTENTIATION IN MANAGEMENT OF MALNUTRITION-INFECTION SYNDROME

Severe malnutrition is complicated by a high mortality rate, mainly from infections, and demands vigorous management. Nutritional supplements, antibiotics, antiparasitic agents, immunization, health education, and improvement in the socioeconomic environment would continue to be the mainstays of the management and prevention of malnutrition-infection syndrome (Chandra, 1977g). Morehead *et al.* (1974) were able to achieve a very high rate of survival by a clear definition of specific infections, isolation of pathogens using aggressive diagnostic procedures (lung puncture, blood culture, tympanocentesis, rectal swabs), and prompt institution of vigorous antibiotic therapy based on probable or actual sensitivity information.

The continuing high mortality of severely malnourished children despite energetic medical management has stimulated attempts at employing additional measures. One such novel approach is immunopotentiation.

An early report (Brown and Katz, 1967) showed that severely malnourished children who were tuberculin negative when first examined on admission into hospital, readily became Mantoux positive after administration of a crude lysate of white cells from a tuberculin positive donor. The subcellular material derived from sensitized lymphocytes that has the capability of transferring to a nonimmune individual the capacity to respond to the antigen is called the "transfer factor." The transfer of sensitivity is usually antigen-specific. The precise nature of the factor is not defined and it is likely that more than one substance is involved in the phenomenon. The dialyzable fraction of transfer factor varies in molecular weight from 700 to 4,000, shows resistance to DNase and RNase, and is heat labile. It contains adenine, cytosine, and guanine. These features fit the properties of double-stranded RNA. The posttransfer ability to respond has been seen within 24-48 hours and is generalized, being evoked at sites far removed from the place of original administration. The recipient is able to provide further material for successive transfer to a nonsensitized individual, which suggests the incorporation of this polynucleotide molecule into a replicating cell. The nondialyzable activity is found in IgG-containing fraction, although the activity does not depend upon the presence of the immunoglobulin molecule. The mechanism(s) of action of the transfer factor is not understood: derepression and adjuvant-like action have been postulated, thereby enhancing immune reactivity. The functional potency of the transfer factor has been employed with partial success in the restoration of cell-mediated immunity in primary immunodeficiency states. Preliminary attempts have been made to exploit these observations by the use of the transfer factor in malnutrition which is consistently associated with secondary defects of cellular immune functions.

Walker *et al.* (1975) conducted a randomized double-blind trial of the effect of the transfer factor in 32 Guatemalan children aged 12 to 48 months recovering from protein–calorie undernutrition. Transfer factor prepared from 500 ml of blood of healthy tuberculin reactor adults from the same community was administered to malnourished children. An unexpected low mortality of 6% precluded the adequate evaluation of the therapy in preventing death associated with nutritional deficiency. However, there was no difference between treated and control groups on several parameters of recovery, including weight gain, hematocrit, reticulocytes, total and individual protein fractions. Therapy with the transfer factor showed no obvious effect on the regular progression of delayed hypersensitivity from almost complete anergy to marked reactivity to tuberculin and *Candida* extract seen in children given nutritional supplements alone. In this study, however, long-term morbidity data were not evaluated. On the other hand, Jose *et al.* (1976), have shown that the administration of lymphocyte transfer factor prepared from a parent's blood was associated with a reduction in diarrheal disease but it afforded no protection against respiratory or skin infections or otitis media. As compared with saline-injected controls, a larger proportion of transfer–factor–treated children maintained or improved in their weight percentiles during the first and second 26-week follow-up periods. Readmission to hospital because of recurrence of infection was similar in the two groups. However in the initial 16 weeks of follow-up, gastroenteritis was not observed in any transfer–factor–treated children. Admissions in treated children were due to pneumonia whereas readmission in control children were mostly with gastroenteritis. The mean duration of hospitalization was shorter and the total loads of fecal parasites tended to be lower in the treated group. The single death in this study occurred in a control child. The mechanism(s) of the apparent beneficial effect of transfer factor in boosting host resistence to diarrheal disease are not clear. In protection against enteropathogens, local mucosal immunity is important. Systemic factors, cellular or humoral, have relatively little role in preventing colonization or infection in mucosal locations. It is unlikely that passive transfer of systemic cellular immunity would enhance resistance to gastroenteritis, as suggested by the authors. Moreover, the lack of data on the distribution of positive fecal cultures among the two groups of patients, and of viral isolations, make the interpretation of results difficult.

Levimasole, an antihelminthic compound, has the ability to stimulate thymus-dependent immunity *in vitro* as well as *in vivo*. Preliminary data suggest that this substance enhances cell-mediated immune response, and reduces the incidence of infections of the respiratory and gastrointestinal tracts (Chandra, 1977b). Prospective studies with careful follow-up observations are planned.

Transfer factor and levimasole are as yet experimental modes of therapy. The two published studies on the effect of transfer factor in malnutrition had different protocols; in one, no beneficial effects were seen whereas in the other, a reduction in diarrhea but not in chest, middle-ear, or skin infections was observed. Levimasole was found to reduce the incidence of infections in a group of undernourished children. More controlled data is required before immunopotentiation can be considered as an adjunct in the therapy of undernutrition.

8.11. OBESITY

Obesity is the most frequent nutritional problem in affluent industrialized countries. Overweight has been correlated with increased mortality, due mainly to atherosclerosis, ischemic heart disease, hypertension, diabetes, and gall bladder disease. Less attention has been given to the increased incidence of infectious disease seen in obese subjects, particularly respiratory infections (Tracey et al., 1971) and postoperative sepsis (Meares, 1975; Printen et al., 1975; Pitkin, 1976). The prevalence of infection as a primary or contributory cause of death is increased in obese individuals (Marks, 1960). Data in animals have been reviewed in Chapter 7.

The mechanisms underlying increased infectious morbidity and mortality in obesity are essentially unknown. Local factors such as reduced vascularity of adipose tissue and restricted pulmonary function are probably important. Data on host defenses in the obese are sadly lacking. Preliminary observations suggest that microbicidal activity of granulocytes is reduced in overweight individuals (Fig. 6.28). The circulating pool and kinetics of PMNs may be altered. The effect of obesity on immunocompetence needs systematic evaluation. It is possible that altered blood lipid and glucose concentrations frequently associated with overnutrition may influence immune responses (Kjösen et al., 1975; Hawley and Gordon, 1976; Waddell et al., 1976; Chandra, 1977b).

The key work for dietary influences to keep immune responses within normal limits is "optimum" nutrition. Too much or too little food is detrimental to health and can menace life.

9

FUTURE RESEARCH NEEDS

The role played by dietary factors in susceptibility to infectious disease has interested research workers for almost a century. Epidemiologic data and observations *in vivo* and *in vitro*, in man and in laboratory animals, have provided the much-needed base. Recent advances in immunologic methodology have led to renewed attempts at solving the maze of nutrition–immunity–infection interactions. The very many lacunae in our knowledge of this subject have been mentioned in earlier sections of this monograph, in the published proceedings of recent conferences (Chandra, 1974b; Katz *et al.*, 1975; Hambraeus *et al.*, 1977; Suskind, 1977), in the comprehensive review of Scrimshaw *et al.* (1968), and in the World Health Organization protocol for field studies and laboratory investigations (1972a). We wish to underscore the urgent need for obtaining information on the following topics:

1. To evaluate the adequacy of humoral, cell-mediated, and secretory immune response to conventional immunization procedures and the effectiveness of vaccine-induced protective immunity.
2. To study the role of total calorie intake and of individual nutrients (protein, amino acids, fats, carbohydrates, vitamins, minerals, trace elements) on immune response and susceptibility to infectious challenge and occurrence of immunopathological disease.

3. Is impaired immunity function in undernutrition a threshold phenomenon or does it alter progressively with increasing severity of deficiency? This question is relevant for postnatal as well as fetal growth retardation associated with nutritional deprivation.
4. To define the role of maternal malnutrition or infection or both in the pathogenesis of low birth weight. This may be achieved by epidemiologic correlative studies or by prophylactic or therapeutic intervention studies.
5. To evaluate the duration of impaired immunocompetence secondary to low birth weight, both preterm and small-for-gestation, and correlate it with the occurrence and magnitude of "catch up" growth.
6. To determine the biologic significance, both short-term and long-term, of altered immune responses of infants suffering from growth retardation due to nutritional deficiencies in the perinatal period or early in childhood.
7. In low-birth-weight infants with pronounced hypoimmunoglobulinemia, to evaluate the effect of administration of pooled human γ-globulin on morbidity due to infection.
8. To define the pathogenetic mechanisms of impaired immune responses in the offspring of nutritionally deprived dams.
9. The effect of infection, its severity and duration, on various aspects of the immune response should be evaluated and analyzed in terms of concomitant nutritional, metabolic, and hormonal changes.
10. What are the effects of microbial products and chalones on host resistance?
11. What is the role of acute-phase reactant proteins in host immune response?
12. To look at the fine structure and histochemical composition of lymphoid tissues in nutritional deprivation.
13. More data are required on the morphological, functional, and circulatory alterations in lymphocyte subpopulations in malnutrition and their clinical and pathogenetic importance.
14. What are the effects of undernutrition on the biochemical and enzymatic activity of various intracellular metabolic pathways?

15. The status of immediate hypersensitivity (Type I) reaction in malnutrition needs study.
16. To determine the carriage rate and types of microbes, including antigens of hepatitis viruses and gut microflora.
17. To develop assessment criteria for diagnosing the presence and severity of nutritional deficiencies. These should preferably be independent of age, sex, ethnic group, and associated infection. The ideal parameter should be simple, nontraumatic, quantitative, reproducible, sensitive, and specific.
18. To develop assessment criteria for the diagnosis of infection. The ideal method should be simple, quantitative, sensitive, and specific.
19. To develop assessment criteria for evaluation of immune responses *in vivo* and *in vitro*. The ideal method should be simple, safe, nontraumatic or use small volumes of blood samples, sensitive, quantitative, and independent of metabolic-hormone homeostatic alterations.
20. To evaluate the role of antigen-nonspecific immunostimulation on immunocompetence and infection-related morbidity and mortality.
21. To study the effect of excess nutrition (obesity, overconsumption of individual nutrients) on immune response and susceptibility to infection in man and in experimental laboratory animals.
22. To determine the effect of hyperlipidemia and individual fats on immunocompetence *in vivo* and *in vitro*.
23. To evaluate critically neutrophil locomotion and ingestion, including the physico-chemical properties and function of actin and myosin, in undernutrition and in obesity.
24. To examine the complete kinetics and pool size of granulocytes.
25. To study macrophage kinetics and function in nutritional deficiency and its importance for the affinity of antibody produced and the occurrence of immunopathological disease.

REFERENCES

Abbassy, A. S., Badr El-Din, M. K., Hassan, A. I., Aref, G. H., Hammad, D. A., El-Araby, I. I., Badr El-Din, A. A., Soliman, M. H., Hussein, M. 1974a. Studies of cell-mediated immunity and allergy in protein energy malnutrition. I. Cell-mediated delayed hypersensitivity. *J. Trop. Med. Hyg.* 77:13.

Abbassy, A. S., Badr El-Din, M. K., Hassan, A. I., Aref, G. H., Hammad, D. A., El-Araby, I. I., and Badr El-Din, A. A. 1974b. Studies of cell-mediated immunity and allergy in protein energy malnutrition. I. Cell-mediated delayed hypersensitivity. *J. Trop. Med. Hyg.* 77:18.

Agnew, L. R. C., and Cook, R. 1949. Antibody production in pyridoxine-deficient rats. *Br. J. Nutr.* 2:321.

Alcantara, E. N., and Speckmann, E. W. 1976. Diet, nutrition and cancer. *Am. J. Clin. Nutr.* 29:1035.

Alexander, J. W., and Wixson, D. 1970. Neutrophil dysfunction and sepsis in burn injury. *Surg. Gynecol. Obstet.* 130:431.

Alford, C. A., Stagro, S., and Reynolds, D. W. 1975. Diagnosis of chronic perinatal infections. *Am. J. Dis. Child.* 129:455.

Altay, C., Say, B., Dogramaci, N., and Bingol, A. 1972. Nitroblue tetrazolium test in children with malnutrition. *J. Pediatr.* 81:392.

Alvarado, J., and Luthringer, D. G. 1971. Serum immunoglobulins in edematous protein–calorie malnourished children. *Clin. Pediatr.* 10:174.

Amin, K., Walia, B. N. S., and Ghai, O. P. 1969. Small bowel function and structure in malnourished children. *Indian Pediatr.* 6:67.

Andelman, M. B., and Sered, B. R. 1966. Utilization of a dietary iron by term infants. A study of 1048 infants from a low socioeconomic population. *Am. J. Dis. Child.* 111:45.

Andersen, V., Hansen, N. E., Karle, H., Lind, I., Hoiby, N., and Weeke, B. 1976. Sequential studies of lymphocyte responsiveness and antibody formation in acute bacterial meningitis, *Clin. Exp. Immunol.* 26:469.

Anderson, J. M. 1972. Increased brain weight/liver weight ratio as a necropsy sign of intrauterine undernutrition. *J. Clin. Pathol.* 25:867.

Anderson, C. G., and Altmann, A. 1951. The electrophoretic serum-protein pattern in malignant malnutrition. *Lancet* 1:203.

Anderson, R., Rabson, A. R., Sher, R., and Koornhof, H. J. 1976. Defective neutrophil motibility in children with measles. *J. Pediatr.* 89:27.

Antia, A. U., McFarlane, H., and Soothill, J. F. 1968. Serum siderophilin in kwashiorkor. *Arch. Dis. Child.* 43:459.

Arbeter, A., Echevarri, L., Fraco, D., Munson, D., Velex, H., and Vitale, J. J. 1971. Nutrition and infection. *Fed. Proc., Fed. Am. Soc. Exp. Biol.* 30:1421.

Aref, G. H., Badr El-Din, M. K., and Hassan, A. I. 1970. Immunoglobulins in kwashiorkor. *J. Trop. Med. Hyg.* 73:186.

Arkwright, J. A., and Zilva, S. S. 1924. Some observations on the effect of diet on the inflammatory reaction. *J. Pathol. Bacteriol.* 27:346.

Aschkenasy, A. 1973. Differing effects of dietary protein deprivation on the production of rosette-forming cells in the lymph nodes and the spleen and on the levels of serum haemagglutinins in rats immunized to sheep red blood cells. *Immunology* 24:617.

Aschkenasy, A. 1974. Effect of a protein free diet on mitotic activity of transplanted splenic lymphocytes. *Nature* 250:325.

Avila, J. L., Valazquez-Avila, G., Correa, C., Castillo, C., and Convit, J. 1973. Leukocytic enzyme differences between the clinical forms of malnutrition. *Clin. Chim. Acta* 19:5.

Axelrod, A. E. 1958. The role of nutritional factors in the antibody responses of the anamnestic process. *Am. J. Clin. Nutr.* 6:119.

Axelrod, A. E., and Hopper, S. 1960. Effects of pantothenic acid, pyridoxine and thiamine deficiencies upon antibody formation to influenza virus PR-8 in rats. *J. Nutr.* 72:325.

Axelrod, A. E., and Trakatellis, A. C. 1964. Induction of tolerance to skin homografts by administering splenic cells to pyridoxine-deficient mice. *Proc. Soc. Exp. Biol. Med.* 116:206.

Axelrod, A. E., Carter, B. B., McCoy, R. H., and Geisinger, R. 1947. Circulating antibodies in vitamin-deficiency states: I: Pyridoxine, riboflavin and pantothenic acid deficiencies. *Proc. Soc. Exp. Biol. Med.* 66:137.

Axelrod, A. E., Fisher, B., Fisher, E., Lee, Y. C. P., and Walsh, P. 1958. Effect of pyridoxine deficiency on skin grafts in the rat. *Science* 127:1388.

Axelrod, A. E., Hopper, S., and Long, D. A. 1961. Effect of pyridoxine deficiency upon circulating antibody formation and skin hypersensitivity reactions to diphtheria toxoid in guinea pigs. *J. Nutr.* 74:58.

Azar, M. M., and Good, R. A. 1971. The inhibitory effect of vitamin A on complement levels and tolerance production. *J. Immunol.* 106:241.

Baehner, R. L., and Nathan, D. G. 1968. Quantitative nitroblue tetrazolium test in chronic granulomatous disease. *N. Engl. J. Med.* 278:971.

Baggs, R. B., and Miller, S. A. 1973. Nutritional iron deficiency as a determinant of host resistance in the rat. *J. Nutr.* 103:1554.

Baggs, R. B., and Miller, S. A. 1974. Defect in resistance to *Salmonella typhimurium* in iron deficient rats. *J. Infect. Dis.* 130:409.

Bainton, D. F., Uliyot, J. L., and Farquhar, M. G. 1971. The development of

REFERENCES

neutrophilic polymorphonuclear leukocytes in human bone marrow. Origin and content of azurophil and specific granules. *J. Exp. Med.* 134:907.

Balch, H. H. 1950. Relation of nutritional deficiency in man to antibody production. *J. Immunol.* 64:397.

Balch, H. H., and Spencer, M. T. 1954. Phagocytosis by human leucocytes. II. Relation of nutritional deficiency in man to phagocytosis. *J. Clin. Invest.* 33:1321.

Bang, B. G., Bang, F. B., and Foard, M. A. 1972. Lymphocyte depression induced in chickens on diets deficient in vitamin A and other components. *Am. J. Pathol.* 68:147.

Bang, F. B., Foard, M., and Bang, B. G. 1973. The effect of vitamin B deficiency in Newcastle disease in lymphoid cells in chickens. *Proc. Soc. Exp. Biol. Med.* 143:1140.

Bang, B. G., Mahalanabis, D., Mukherjee, K. L., and Bang, F. B. 1975. T and B lymphocyte rosetting in undernourished children. *Proc. Soc. Exp. Biol. Med.* 149:199.

Banihashemi, A., Kohout, E., Dutz, W., and Rafii, R. 1973. Abdominal lymphoma and abnormal proteins in alpha heavy chain disease in Iran. *Pahlavi Med. J.* 4:377.

Basta, S. S., and Churchill, A. 1974. "Iron Deficiency Anemia and the Productivity of Adult Males in Indonesia." Staff Working Paper No. 175, World Bank, Washington, D.C.

Bean, W. B., and Hodges, R. E. 1954. Pantothenic acid deficiency induced in human subjects. *Proc. Soc. Exp. Biol. Med.* 86:693.

Beisel, W. R. 1972. Interrelated changes in host metabolism during generalized infectious illness. *Am. J. Clin. Nutr.* 25:1254.

Beisel, W. R. 1975. Metabolic response to infection. *Annu. Rev. Med.* 26:9.

Beisel, W. R. 1977. Malnutrition as a consequence of stress, *in* "Malnutrition and the Immune Response" (R. M. Suskind, ed.), Raven Press, New York, p. 21.

Bell, R. G., and Hazell, L. A. 1975. Influence of dietary protein restriction on immune competence. 1. Effect on the capacity of cells from various lymphoid organs to induce graft-vs.-host reactions. *J. Exp. Med.* 141:127.

Bell, R. G., Turner, K. J., Gracey, M., Suharjano, and Sunoto. 1976. Serum and small intestinal immunoglobulin levels in undernourished children. *Am. J. Clin. Nutr.* 29:392.

Bellanti, J. A. 1970. "Immunology," Saunders, Philadelphia.

Bellanti, V. A., and Dayton, D. H., eds. 1975. "The Phagocytic Cell in Host Resistance," Raven Press, New York.

Bellanti, J. A., Krasner, R., Bartelloni, P. J., Yang, M. C., and Beisel, W. R. 1972. Sandfly fever: Sequential changes in neutrophil biochemical and bactericidal functions. *J. Immunol.* 108:142.

Benditt, E. P., Wissler, R. W., Woolridge, R. L., Rowley, D. A., and Steffee, C. H. 1949. Loss of body protein and antibody production by rats on low protein diets. *Proc. Soc. Exp. Biol.* 70:240.

Bengoa, J. J. 1974. The problem of malnutrition. *WHO Chron.* 28:3.

Berenbaum, M. C. 1975. The clinical pharmacology of immunosuppressive agents, in "Clinical Aspects of Immunology" (P. G. H. Gell, R. R. A. Coombs, and P. J. Lachmann, eds.), Blackwell, Oxford, p. 689.

Berlin, R. D., Oliver, J. M., Ukena, T. E., and Yin, H. H. 1975. The cell surface. *New Engl. J. Med.* 292:515.

Berry, L. J., Davis, J., and Spies, T. D. 1945. The relationship between diet and the mechanism for defense against bacterial infections in rats. *J. Lab. Clin. Med.* 30:684.

Bhaskaram, C., and Reddy, V. 1975. Cell-mediated immunity in iron- and vitamin-deficient children. *Br. Med. J.* 3:522.

Bhuyan, U. N., and Ramalingaswami, V. 1972. Responses of the protein-deficient rabbit to staphylococcal bacteremia. *Am. J. Pathol.* 69:359.

Bhuyan, U. N., and Ramalingaswami, V. 1973. Immune responses of the protein-deficient guinea pig of BCG vaccination. *Am. J. Pathol.* 72:489.

Bhuyan, U. N., and Ramalingaswami, V. 1974. Lymphopoiesis in protein deficiency. *Am. J. Pathol.* 75:315.

Bhuyan, U. N., Mohapatra, L. N., and Ramalingaswami, V. 1974. Phagocytosis, bactericidal activity and nitroblue tetrazolium reduction by the rabbit neutrophil in protein malnutrition. *Indian J. Med. Res.* 62:42.

Bieling, R. 1925. Die Wirkung von Bakteriengiften auf unterernährte Tiere (Tuberkulin, Diphtherie-toxin). *Z. Hyg. Infektionskr.* 104:518.

Bistrian, B. R., Blackburn, G. L., Scrimshaw, N. S., and Flatt, J. P. 1975. Cellular immunity in semistarved states in hospitalized adults. *Am. J. Clin. Nutr.* 28:1148.

Blackberg, S. M. 1927–28. Effect of the immunity mechanism of various avitaminoses. *Proc. Soc. Exp. Biol. Med.* 25:770.

Borum, K. 1972. Effect of neonatal thymectomy on the primary hemolysin response and on lymph node cell count in five strains of mice. *Acta Pathol. Microbiol. Scand.* 80:287.

Bradford, W. L. 1928. Mucosus organism from suppurative lesions of rats on diet deficient in Vitamin A. *J. Infect. Dis.* 43:407.

Brody, J. A., Overfield, T., and Hammes, L. M. 1964. Depression of the tuberculin reaction by viral vaccines. *N. Engl. J. Med.* 271:1294.

Brown, R. E., and Katz, M. 1965. Antigen stimulation in undernourished children. *E. Afr. Med. J.* 42:221.

Brown, R. E., and Katz, M. 1966. Failure of antibody production to yellow-fever vaccine in children with kwashiorkor. *Trop. Geogr. Med.* 18:125.

Brown, R., and Katz, M. 1967. Passive transfer of delayed hypersensitivity reaction to tuberculin in children with protein calorie malnutrition. *J. Pediatr.* 70:126.

Buchanan, B. J., and Filkins, J. P. 1976. Hypoglycemic depression of RES function. *Am. J. Physiol.* 231:265.

Bullen, J. J., Rogers, H. J., and Leigh, L. 1972. Iron-binding proteins in milk and resistance to Escherichia coli infection in infants. *Br. Med. J.* 1:69.

Burch, P. R. J. 1968. "An Inquiry Concerning Growth, Disease and Ageing," Oliver and Boyd, Edinburgh.

Burgess, B. J., Vos, G. H., Coovadia, H. M., Smythe, P. M., Parent, M. A., and Loening, W. E. L. 1974. Radio-isotope assessment of phytohaemagglutinin-stimulated lymphocytes from patients with protein calorie malnutrition. *S. Afr. Med. J.* 48:1870.

Burkitt, D. P. 1971. Epidemiology of cancer of the colon and rectum. *Cancer* 28:3.

Burman, D. 1972. Hemoglobin levels in normal infants aged 3–24 months and the effect of iron. *Arch. Dis. Child.* 47:261.

Burnet, F. M. 1970. "Immunological Surveillance," Pergamon Press, New York.

Burnet, F. M., ed. 1976. "Immunology," Freeman, San Francisco.

Bwibo, N. O., and Owor, R. 1970. *Pneumocystis carinii* pneumonia in Uganda African children. *West Afr. Med. J.* 19:184.

Cameron, E., and Pauling, L. 1973. Ascorbic acid and the glycosaminoglycans. An orthomolecular approach to cancer and other diseases. *Oncology* 27:181.

Cameron, E., and Pauling, L. 1974. The orthomolecular treatment of cancer. I. The role of ascorbic acid in host resistance. *Chem. Biol. Interact.* 9:273.

Campbell, S. J. 1970. Ultrasonic fetal cephalometry during the second trimester of pregnancy. *J. Obstet. Gynaecol. Br. Commonw.* 77:1057.

Cannon, P. R., Chase, W. E., and Wissler, R. W. 1943. The relationship of protein reserves to antibody production. I. The effects of low-protein diet and of plasmapheresis upon the formation of agglutinins. *J. Immunol.* 47:133.

Canonico, P. G., White, J. D., and Powanda, M. C. 1975. Peroxisome depletion in rat liver during pneumococcal sepsis. *Lab. Invest.* 33:147.

Carr, I. 1973. "The Macrophage: A Review of Ultrastructure and Function," Academic Press, New York.

Chanarin, I., Rothman, D., Ward, A., and Perry, J. 1968. Folate status and requirement in pregnancy. *Br. Med. J.* 2:390–394.

Chandler, A. C., Read, C. P., and Nicholas, H. O. 1950. Observations on certain phases of nutrition and host–parasite relations of Hymenolepis diminuta in white rats. *J. Parasitol.* 32:523.

Chandra, R. K. 1970. Immunological picture in Indian childhood cirrhosis. *Lancet* 1:537.

Chandra, R. K. 1972. Immunocompetence in undernutrition. *J. Pediatr.* 81:1184.

Chandra, R. K. 1973a. Reduced bactericidal capacity of polymorphs in iron deficiency. *Arch. Dis. Child.* 48:863.

Chandra, R. K. 1973b. Polymorph dysfunction in nutritional deficiency states *in* "Malnutrition and Function of Blood Cells" (N. Shimazone and T. Arakawa, eds.), Japanese Panel of Malnutrition, National Institute of Nutrition, Tokyo, p. 333.

Chandra, R. K. 1974a. Rosette forming T lymphocytes and cell-mediated immunity in malnutrition. *Br. Med. J.* 3:608.

Chandra, R. K. 1974b. Interactions of infection and malnutrition *in* "Progress in Immunology II" (L. Brent and J. Holborow, eds.), North Holland, Amsterdam, vol. 4, p. 355.

Chandra, R. K. 1974c. Immunocompetence in low-birth-weight infants after intrauterine malnutrition. *Lancet,* 2:1393.

Chandra, R. K. 1975a. Serum complement and immunoconglutinin in malnutrition. *Arch. Dis. Child.* 50:225.

Chandra, R. K. 1975b. Levels of IgA subclasses, IgA, IgM and tetanus antitoxin in paired maternal and foetal sera: Findings in healthy pregnancy and placental insufficiency *in* "Materno-foetal Transmission of Immunoglobulins" (W. A. Hemmings, ed.), Cambridge University Press, Cambridge, p. 77.

Chandra, R. K. 1975c. Fetal malnutrition and postnatal immunocompetence. *Am. J. Dis. Child.* 129:450.

Chandra, R. K. 1975d. Impaired immunocompetence associated with iron deficiency. *J. Pediatr.* 86:899.

Chandra, R. K. 1975e. Food antibodies in malnutrition. *Arch. Dis. Child.* 50:532.

Chandra, R. K. 1975f. Reduced serum and secretory antibody response to live attenuated measles and poliovirus vaccines in malnourished children. *Br. Med. J.* 2:583.

Chandra, R. K. 1975g. Antibody formation in first and second generation offspring of nutritionally deprived rats. *Science* 190:289.

Chandra, R. K. 1975h. T-lymphocytes and cell-mediated immunity in low-birth-weight infants. *Pediatr. Res.* 9:328.

Chandra, R. K. 1975i. Lymphocyte response to hepatitis B surface antigen: Findings in hepatitis and Indian childhood cirrhosis. *Arch. Dis. Child.* 50:559.

Chandra, R. K. 1976a. Iron and immunocompetence. *Nutr. Rev.* 34:129.

Chandra, R. K. 1976b. Nutrition as a critical determinant in susceptibility to infection. *World Rev. Nutr. Diet.* 25:166.

Chandra, R. K. 1976c. Nutrition and leukocyte function *in* "Proceedings Tenth International Congress of Nutrition," Kyoto, Japan, p. 320.

Chandra, R. K. 1976d. Iron-deficiency anemia and immune responses. *Lancet* 2:1200.

Chandra, R. K. 1976e. Separation and functional characteristics of lymphocyte subpopulations in peripheral blood. *Pediatr. Res.* 10:384.

Chandra, R. K. 1976f. Indian childhood cirrhosis: Genealogic data, alpha-fetoprotein, hepatitis antigen and circulating immune complexes. *Trans. R. Soc. Trop. Med. Hyg.* 70:296.

Chandra, R. K. 1977a. Lymphocyte subpopulations in malnutrition: Cytotoxic and suppressor cells. *Pediatrics* 59:423.

Chandra, R. K. 1977b. In preparation.

Chandra, R. K. 1977c. Immunoglobulins and antibody response in malnutrition *in* "Malnutrition and the Immune Response" (R. Suskind, ed.), Raven Press, New York, p. 155.

Chandra, R. K. 1977d. Cell-mediated immunity in fetally and postnatally malnourished children from India and Newfoundland *in* "Malnutrition and the Immune Response" (R. Suskind, ed.), Raven Press, New York, p. 111.

Chandra, R. K. 1977e. Serum complement components in malnourished Indian children *in* "Malnutrition and the Immune Response" (R. Suskind, ed.), Raven Press, New York, p. 329.

Chandra, R. K. 1977f. Immunological consequences of malnutrition including fetal growth retardation *in* "Food and Immunology" (L. Hambraeus, L. Å. Hanson, and H. McFarlane, eds.), Swedish Nutrition Foundation Symposium XIII, Almqvist and Wiksell International, Stockholm, p. 58.

Chandra, R. K. 1977g. Nutritional modulation of immune response *in* "Practice of Pediatrics" (V. C. Kelley, ed.), Harper and Row, New York.

Chandra, R. K. 1977h. Energy-protein undernutrition *in* "Practice of Pediatrics" (V. C. Kelley, ed.), Harper and Row, New York.

Chandra, R. K. 1977i. Unpublished data.

Chandra, R. K. 1978a. Interactions of nutrition, infection and immune response. *Acta Paediatr. Scand.*, in press.

Chandra, R. K. 1978b. Ontogenetic development of immune system and effects of fetal growth retardation. *J. Perinat. Med.*, in press.

Chandra, R. K. 1978c. The influence of nutritional status on susceptibility to infection. *Adv. Nutr. Res.*, in press.

Chandra, R. K., and Bhujwala, R. A. 1977. Elevated α-fetoprotein and impaired immune response in malnutrition. *Int. Arch. Allergy Appl. Immunol.* 53:80.

Chandra, R. K., and Ghai, O. P. 1972. Serum immunoglobulins in healthy children from birth to adolescence. *Indian J. Med. Res.* 60:89.

Chandra, R. K., Guha, D. K., and Ghai, O. P. 1970. Serum immunoglobulins in the newborn. *Indian J. Pediatr.* 37:361.

Chandra, R. K., Sharma, S., and Bhujwala, R. A. 1973. Effect of acute and chronic starvation on plaque forming cell response in mice. *Indian J. Med. Res.* 61:93.

Chandra, R. K., Chandra, S., and Ghai, O. P. 1976a. Chemotaxis, random mobility and mobilization of polymorphonuclear leucocytes in malnutrition. *J. Clin. Pathol.* 29:224.

Chandra, R. K., Chakraburty, S., and Chandra, S. 1976b. Malnutrition, humoral immunity and infection. *Indian J. Pediatr.* 43:159.

Chandra, R. K., Bhujwala, R. A., and Chandra, S. 1976c. Immunoregulatory role of alpha-fetoprotein. *Pediatr. Res.* 10:384.

Chandra, R. K., Seth, V., Chandra, S., Bhujwala, R. A., and Ghai, O. P. 1977a. Polymorphonuclear leukocyte function in malnourished Indian children *in* "Malnutrition and the Immune Response" (R. Suskind, ed.), Raven Press, New York, p. 259.

Chandra, R. K., Khalil, N., Howse, D., Chandra, S., and Kutty, K. M. 1977b. Lysozyme (muramidase) activity in plasma, neutrophils and urine in malnutrition and infection *in* "Malnutrition and the Immune Response" (R. Suskind, ed.), Raven Press, New York, p. 407.

Chandra, R. K., Ali, S. K., Kutty, K. M., and Chandra, S. 1977c. Thymus-dependent lymphocytes and delayed hypersensitivity in low birth weight infants. *Biol. Neonat.* 31:15.

Chandra, R. K., Au, B., Woodford, G., and Hyam, P. 1977d. Iron status, immunocompetence and susceptibility to infection *in* "Ciba Foundation Symposium on Iron Metabolism" (A. Jacob, ed.), Elsevier, Amsterdam, p. 249.

Cheson, B. D., Curnutte, J. T., and Babior, B. M. 1977. The oxidative killing mechanisms of the neutrophil *in* "Progress in Clinical Immunology" (R. S. Schwartz, ed.), Grune and Stratton, New York, vol. 3, p. 9.

Claman, H. N. 1972. Corticosteroids and lymphoid cells. *N. Engl. J. Med.* 287:388.

Claphan, P. A. 1934. Ascariasis and vitamin A deficiency in pigs. *J. Helminthol.* 12:165.

Clausen, S. W. 1935. Nutrition and infection. *J. Am. Med. Assoc.* 104:793.

Clayson, D. B. 1975. Nutrition and experimental carcinogenesis: A review. *Cancer Res.* 35:3292.

Cline, M. J. 1975. "The White Cell." Boston University Press, Boston.

Cohen, S., and Hansen, J. D. L. 1962. Metabolism of albumin and γ-globulin in kwashiorkor. *Clin. Sci.* 23:351.

Cohen, S., Metz, J., and Hartz, D. 1962. Protein-losing gastroenteropathy in kwashiorkor. *Lancet* 1:52.

Colten, H. R. 1976. Biosynthesis of complement. *Adv. Immunol.* 22:67.

Cook, G. C. 1972. Impairment of d-xylose absorption in Zambian patients with systemic bacterial infections. *Am. J. Clin. Nutr.* 25:490.

Coombs, R. R. A., and Gell, P. G. H. 1975. Classification of allergic reactions responsible for clinical hypersensitivity and disease *in* "Clinical Aspects of Immunology" (P. G. H. Gell, R. R. A. Coombs, and P. J. Lachman, eds.), Blackwell, Oxford, p. 761.

Coombs, R. R. A., and Smith, H. 1975. The allergic response and immunity *in* "Clinical Aspects of Immunology" (P. G. H. Gell, R. R. A. Coombs, and P. J. Lachmann, eds.), Blackwell, Oxford, p. 473.

Cooper, M. D., Perey, D. Y., Gabrielsen, A. R., Sutherland, D. E. R., McKneally, M. F., and Good, R. A. 1968. Production of an antibody deficiency syndrome in rabbits by neonatal removal of organized intestinal lymphoid tissues. *Int. Arch. Allergy Appl. Immunol.* 33:65.

Cooper, M. D., Lawton, A. R., and Kincade, P. W. 1972. A two stage model for development of antibody-producing cells. *Clin. Exp. Immunol.* 11:143.

Cooper, W. C., Mariani, T., and Good, R. A. 1970. Effects of chronic protein depletion on immune response. *Fed. Proc., Fed. Am. Soc. Exp. Biol.* 29:364.

Cooper, W. C., Good, R. A., and Mariani, T. 1974. Effects of protein insufficiency on immune responsiveness. *Am. J. Clin. Nutr.* 27:627.

Coovadia, H. M., and Soothill, J. F. 1976a. The effect of protein restricted diets on the clearance of ^{125}I-labelled polyvinyl pyrrolidone in mice. *Clin. Exp. Immunol.* 23:373.

Coovadia, H. M., and Soothill, J. F. 1976b. The effect of amino acid restricted diets on the clearance of ^{125}I-labelled polyvinyl pyrrolidone in mice. *Clin. Exp. Immunol.* 23:562.

REFERENCES

Coovadia, H. M., Parent, M. A., Loening, W. E. K., Wesley, A., Burgess, B., Hallett, D., Brain, P., Grace, J., Naidoo, J., Smythe, P. M., and Vos, G. H. 1974. An evaluation of factors associated with the depression of immunity in malnutrition and in measles. *Am. J. Clin. Nutr.* 27:665.

Coovadia, H. M., Wesley, A., Brain, P., Henderson, L. G., Hallett, A. F., and Vos, G. H. 1977. Immunoparesis and outcome in measles. *Lancet* 1:619.

Cort, W. W., and Otto, G. F. 1940. Immunity and hook worm disease. *Rev. Gastroenterol.* 7:2.

Craddock, P. R., Yawata, Y., Van Santen, L., Gilberstadt, S., Silvis, S., and Jacob, H. S. 1974. Acquired phagocyte dysfunction: A complication of the hypophosphatemia of parenteral hyperalimentation. *N. Engl. J. Med.* 290:1403.

Craft, A. W., Reid, M. M., and Low, W. T. 1976. Effect of virus infection on polymorph function in children. *Lancet*, 1:1570.

Crane, C. S. 1965. Infectious bovine rhinotracheitis abortion and its relationship to nutrition in California beef cattle. *J. Am. Vet. Med. Assoc.* 147:1308.

Crawford, M. A. 1968. Food selection under natural conditions and the possible relationship to heart disease in man. *Proc. Nutr. Soc.* 27:163.

Cunningham, A. S. 1976. Lymphomas and animal-protein consumption. *Lancet* 2:1184.

Curry, C. R., and Quie, P. G. 1971. Fungal septicemia in patients receiving parenteral hyperalimentation. *N. Engl. J. Med.* 205:1221.

Dally, P. 1969. "Anorexia Nervosa," Heinemann, London.

Davis, D. S., Nelson, T., and Shepard, T. H. 1970. Teratogenicity of vitamin D_3 deficiency: Omphalocele, skeletal and neural defects and splenic hypoplasia. *Science* 169:1329.

Demarchi, M. 1958. Effect of dietary protein on blood regeneration of anemic patients suffering from parasite infestations. *Am. J. Clin. Nutr.* 6:415.

Deo, M. G., Bhan, I., and Ramalingaswami, V. 1973. Influence of dietary protein deficiency on phagocytic activity of the reticulo-endothelial cells. *J. Pathol.* 109:215.

dePablo, D. L., Ramirez, A., and Kumate, J. 1972. Graft versus host reactions in severe malnutrition of mice and chickens *in* "Proceedings of the Ninth International Congress on Nutrition, Mexico," Karger, Basel, vol. 12, p. 155.

Department of Health, Education, and Welfare. 1972. The ten-state nutrition survey, 1968-1970. Publication No. (HSM) 72-8130 to 72-8134. Center for Disease Control, Atlanta.

Dossetor, J. F. B., and Whittle, H. C. 1975. Protein-losing enteropathy and malabsorption in acute measles enteritis. *Br. Med. J.* 2:592.

Douglas, S. D., and Schopfer, K. 1974. Phagocyte function in protein calorie malnutrition. *Clin. Exp. Immunol.* 17:121.

DuBois, R. J. 1955. The effect of metabolic factors on susceptibility of albino mice to experimental tuberculosis. *J. Exp. Med.* 101:58.

DuBois, R. J., and Schaedler, R. W. 1958. Effect of dietary proteins and amino acids on the susceptibility of mice to bacterial infections. *J. Exp. Med.* 108:69.

Duncan, P. G., and Cullen, B. F. 1976. Anesthesia and immunology. *Anesthesiology* 45:522.
Dutz, W. 1970. Pneumocystis carinii pneumonia. *Pathobiol. Annu.* 5:309.
Dutz, W. 1975. Immune modulation and disease patterns in population groups. *Medical Hypotheses* 5:197.
Dutz, W., Asvadi, S., Sadri, S., and Kohout, E. 1971. Intestinal lymphoma and sprue: A systematic approach. *Gut* 12:804.
Dutz, W., Rossipal, E., Ghavami, H., Vessel, K., Kohout, E., and Post, G. 1976. Persistent cell mediated immune deficiency following infantile stress during the first 6 months of life. *Eur. J. Pediatr.* 122:117.
Dwyer, J. M., and Kantor, F. S. 1973. Regulation of delayed hypersensitivity: Failure to transfer delayed hypersensitivity to desensitized guinea pigs. *J. Exp. Med.* 137:32.
Edelman, R., Suskind, R., Olson, R. E., and Sirisinha, S. 1973. Mechanisms of defective delayed cutaneous hypersensitivity in children with protein-calorie malnutrition. *Lancet* 1:506.
Ederer, F., Leren, P., Turpeinen, O., and Frantz, I. D. 1971. Cancer among men on cholesterol-lowering diets. *Lancet* 2:203.
El Gholmy, A., Hashish, S., Helmy, O., Aly, R. H., and E. Gomal, Y. 1970. A study of immunoglobulins in kwashiorkor and marasmus. *J. Trop. Med. Hyg.* 73:192.
Elsbach, P. 1973. On the interactions between phagocytes and micro-organisms. *N. Engl. J. Med.* 289:846.
Eriksson, B., and Hedfors, E. 1977. The effect of adrenaline, insulin and hydrocortisone on human peripheral blood lymphocytes studied by cell surface markers. *Scand. J. Haematol.* 18:121.
FAO/WHO Expert Committee on Nutrition. 1971. Eighth Report. WHO Tech. Rep. Ser. 477.
Farid, N. R., Au, B., Woodford, G., and Chandra, R. K. 1976. Polymorphonuclear leucocyte function in hypothyroidism. *Hormone Res.* 7:247.
Faulk, W. P., and Chandra, R. K. 1977. Nutrition and resistance *in* "Hand Book of Nutrition" (M. Recheigl, ed.), CRC Press, Cleveland.
Fearon, D. T., and Austen, K. F. 1975. Properdin: Initiation of alternative complement pathway. *Proc. Nat. Acad. Sci. U.S.A.* 72:3220.
Feldman, G., and Gianantonio, C. A. 1972. Aspectos immunologicos de la desnutricion en el niño, *Medicina* 32:1.
Felig, P., Brown, W. V., Levine, R. A., and Klatskin, G. 1970. Glucose homeostasis in viral hepatitis. *N. Engl. J. Med.* 283:1436.
Ferguson, A. S., Lawlor, G. J., Jr., Neumann, C. G., Oh, W., and Stiehm, E. R. 1974. Decreased rosette-forming lymphocytes in malnutrition and intrauterine growth retardation. *J. Pediatr.* 85:717.
Fernandes, G., Yunis, E. J., and Good, R. A. 1976. Influence of protein restriction on immune functions in NZB mice. *J. Immunol.* 116:782.
Finkelstein, M. H. 1931–32. Effect of carotene on course of M. tuberculosis infection of mice fed vitamin A deficiency diet. *Proc. Soc. Exp. Biol. Med.* 21:969.

REFERENCES

Fiser, R. H., Rollins, J. B., and Beisel, W. R. 1972. Depressed resistance to infectious canine hepatitis in dogs on a high fat diet. *Am. J. Vet. Res.* 33:713.

Fletcher, J., Mather, J., Lewis, M. J., and Whiting, G. 1975. Mouth lesions in iron-deficient anemia: Relationship to candida albicans in saliva and to impairment of lymphocyte transformation. *J. Infect. Dis.* 131:44.

Foster, A. O. 1936. On a probable relationship between anemia and susceptibility to hookworm infection. *Am. J. Hyg.* 24:109.

Foster, C., Jones, J. H., Henle, W., and Brenner, S. A. 1949. Nutrition and poliomyelitis: The effects of deficiencies of phosphorus, calcium and vitamin D on the response of mice to the Lansing strain of poliomyelitis virus. *J. Infect. Dis.* 85:173.

Freyre, E. A., Chabes, A., Poemape, O., and Chages, A. 1973. Abnormal Rebuck skin-window response in kwashiorkor. *J. Pediatr.* 82:523.

Gaafar, S. M., and Ackert, J. E. 1958. Studies on mineral deficient diets as factors in resistance of fowls to parasitism. *Exp. Parasitol.* 2:185.

Gautam, S. C., Aikat, B. K., and Sehgal, S. 1973. Immunological studies in protein malnutrition I. Humoral and cell-mediated immune response in protein-deficient mice. *Indian J. Med. Res.* 61:78.

Gay, W. I. 1968. "Methods of Animal Experimentation," vol. 3, Academic Press, New York.

Gebhardt, B. M., and Newberne, P. M. 1974. Nutrition and immunological responsiveness. T-cell function in the offspring of lipotrope and protein-deficient rats. *Immunology* 26:489.

Geefhuysen, J., Rosen, E. U., Katz, J., Ipp, T., and Metz, J. 1971. Impaired cellular immunity in kwashiorkor with improvement after therapy. *Br. Med. J.* 4:527.

Gell, P. G. H. 1948. Discussion on nutrition and resistance to infection. *Proc. R. Soc. Med.* 41:323.

Gell, P. G. H., Coombs, R. R. A., and Lachman, P. J., eds., 1975. "Clinical Aspects of Immunology," Blackwell, Oxford.

Gemeroy, D. G., and Koffler, A. H. 1949. The production of antibodies in protein depleted and repleted rabbits. *J. Nutr.* 39:299.

Getz, H. R., Long, E. R., and Henderson, H. J. 1951. A study on the relation of nutrition to the development of tuberculosis. Influence of ascorbic acid and vitamin A. *Am. Rev. Tuberc.* 64:381.

Ghai, O. P., and Jaiswal, V. N. 1970. Relationship of undernutrition to diarrhoea in infants and children. *Indian J. Med. Res.* 58:789.

Ghosh, S., and Dhatt, P. S. 1961. Complications of measles. *Indian J. Child Health* 10:111.

Ghosh, S., Kumari, S., Balaya, S., and Bhargava, S. K. 1970. Antibody response to oral polio vaccine in infancy. *Indian Pediatr.* 7:78.

Giron-Mendez, R. A. 1963. "Reaccion de dos nematodos intestinalli al cambio de dieta en perros." Thesis, Guatemala, Escuela Nacional Central de Agricultura.

Gitlin, D., and Biasucci, A. 1969. Development of γG, γA, γM, $\beta_{1C/I_{A'}}$, C' esterase inhibitor, ceruloplasmin, haptoglobin, fibrinogen, plasminogen, α_1-antitrypsin, orosomucoid, β-lipoprotein, α_2-macroglobulin and prealbumin in the human conceptus. *J. Clin. Invest.* 48:1433.

Goldberg, A. 1959. The relationship of diet to gastrointestinal parasitism in cattle. *Am. J. Vet. Res.* 20:806.

Gomez, F., Ramos-Galvan, R., Frenk, S., Cravioto, J. M., Chavez, R., and Vasquez, J. 1956. Mortality in third degree malnutrition. *J. Trop. Pediatr.* 2:77.

Good, R. A. 1973. Crucial experiments of Nature that have guided analysis of the immunologic apparatus *in* "Immunologic Disorders in Infants and Children" (E. R. Stiehm and V. A. Fulginiti, eds.), Saunders, Philadelphia, p. 3.

Good, R. A., Fernandes, G., Yunis, E. J., Cooper, W. C., Jose, D. C., Kramer, T. R., and Hansen, M. A. 1976. Nutritional deficiency, immunologic function and disease. *Am. J. Pathol.* 84:599.

Good, R. A., Jose, D., Cooper, W. C., Fernandes, G., Kramer, T., and Yunis, E. 1977. Influence of nutrition on antibody production and cellular immune responses in man, rats, mice and guinea pigs *in* "Malnutrition and the Immune Responses" (R. M. Suskind, ed.), Raven Press, New York, p. 169.

Gordon, J. E., Jansen, A. A. J., and Ascoli, W. 1965. Measles in rural Guatemala. *J. Pediatr.* 66:779.

Gordon, Y. B., Grudzinskas, J. G., Jeffrey, D., Chard, T., and Letchworth, A. T. 1977. Concentration of pregnancy-specific B_1-glycoprotein in maternal blood in normal pregnancy and in intrauterine growth retardation. *Lancet* 1:331.

Gotch, F. M., Spry, C. J. F., Mowat, A. G., Beeson, P. B., and MacLennan, I. C. M. 1975. Reversible granulocyte killing defect in anorexia nervosa. *Clin. Exp. Immunol.* 21:244.

Grace, H. J., Armstrong, D., and Smythe, P. M. 1972. Reduced lymphocyte transformation in protein-calorie malnutrition. *S. Afr. Med. J.* 46:402.

Gracey, M., and Stone, D. E. 1972. Small intestinal microflora in Australian aboriginal children with chronic diarrhoea. *Aust. N.Z. J. Med.* 2:215.

Gratzl, E., Hromatka, L., and Ullrich, W. 1963. Ein Beitrag zur Bekämpfung der infektiven Bronchitis der Hühner. *Wein Tierarztl. Wochenschr.* 50:37.

Greenberg, L. J., and Yunis, E. J. 1972. Immunologic control of aging: A possible primary event. *Gerontologia* 18:247.

Greene, M. R. 1933. The effects of vitamins A and D on antibody production and resistance to infection. *Am. J. Hyg.* 17:60.

Gross, R. L., Reid, J. V. O., Newberne, P. M., Burgess, B., Marston, R., and Hift, W. 1975. Depressed cell-mediated immunity in megaloblastic anemia due to folic acid deficiency. *Am. J. Clin. Nutr.* 28:225.

Grossberg, S. E. 1972. The interferons and their inducers: Molecular and therapeutic considerations. *N. Engl. J. Med.* 287:13, 79, 122.

Grove, D. I., Burston, T. O., and Forbes, I. J. 1974. Immunoglobin E and eosinophil levels in atopic and nonatopic populations infested with hookworm. *Clin. Allergy* 4:295.

REFERENCES

Gruenwald, P., ed., 1975. "The Placenta," Medical Technical Publishing, Lancaster.
Guggenheim, K., and Buechler, E. 1947. Nutritional deficiency and resistance to infection. *Hygiene* 45:103.
Haider, S., de Coutinho, M., Emond, R. T. D., and Sutton, R. N. P. 1973. Tuberculin anergy and infectious mononucleosis. *Lancet* 2:74.
Hambraeus, L., Hanson, L. Å., and McFarlane, H., eds., 1977. "Food and Immunology," Almqvist and Wiksell, Stockholm.
Hansen, N. E., and Andersen, V. 1973. Lysozyme activity in human neutrophilic granulocytes. *Br. J. Haematol.* 24:607.
Hansen, N. E., Karle, H., Andersen, V., Malmquist, J., and Hoff, G. E. 1976. Neutrophilic granulocytes in acute bacterial infection. Sequential studies on lysozyme, myeloperoxidase and lactoferrin. *Clin. Exp. Immunol.* 26:463.
Hanson, L. Å., and Brandtzaeg, P. 1973. Secretory antibody systems *in* "Immunologic Disorders in Infants and Children" (E. R. Stiehm and V. A. Fulginiti, eds.), Saunders, Philadelphia, p. 107.
Harland, P. S. E. G. 1965. Tuberculin reactions in malnourished children. *Lancet* 1:719.
Harmon, B. G., Miller, E. R., Hoefer, J. A., Ullrey, D. E., and Luecke, R. W. 1963. Relationship of specific nutrient deficiencies to antibody production in swine. *J. Nutr.* 79:263.
Harrison, B. D. W., Tugwell, P., and Fawcett, I. W. 1975. Tuberculin reaction in adult Nigerians with sputum-positive pulmonary tuberculosis. *Lancet* 1:421.
Hawley, H. P., and Gordon, G. B. 1976. The effects of long-chain free fatty acids on human neutrophil function and structure. *Lab. Invest.* 34:216.
Hedgecock, L. W. 1958. The effect of diet on the inducement of acquired resistance by viable and nonviable vaccines in experimental tuberculosis. *Am. Rev. Tuberc.* 77:93.
Heiss, L. I., and Palmer, D. L. 1974. Anergy in patients with leukocytosis. *Am. J. Med.* 56:323.
Hershko, C., Karsai, A., Eylon, L., and Izak, G. 1970. The effect of chronic iron deficiency on some biochemical functions of the human hemopoietic tissue. *Blood* 36:321.
Heyworth, B., and Brown, J. 1975. Jejunal microflora in malnourished Gambian children. *Arch. Dis. Child.* 50:27.
Higashi, O., Sato, Y., Takamatsu, H., and Oyama, M. 1967. Mean cellular peroxidase (MCP) of leukocytes in iron deficiency anemia. *Tohoku J. Exp. Med.* 93:105.
Higgs, J. M., and Wells, R. S. 1973. Chronic mucocutaneous candidiasis: Associated abnormalities of iron metabolism. *Br. J. Dermatol., Suppl.* 8:88.
Hill, G. L., Blackett, R. L., Pickford, I., Burkinshaw, L., Young, G. A., Warren, J. V., Schorah, C. J., and Morgan, D. B. 1977. Malnutrition in surgical patients: An unrecognized problem. *Lancet* 1:689.
Hiraishi, T. 1927. Experimental ascariasis of the young pigs with special reference to A vitaminosis. *J. Keio. Med. Soc.* (Abstr.) in *Jpn. Med. World* 7:79.

Hodges, R. E., Bean, W. B., Ohlson, M., and Bleiler, R. E. 1962a. Factors affecting human antibody response. III. Immunologic responses of men deficient in pantothenic acid. *Am. J. Clin. Nutr.* 11:85.

Hodges, R. E., Bean, W. B., Ohlson, M. A., and Bleiler, R. E. 1962b. Factors affecting human antibody response. IV. Pyridoxine deficiency. *Am. J. Clin. Nutr.* 11:180.

Hodges, R. E., Bean, W. B., Ohlson, M. A., and Bleiler, R. E. 1962c. Factors affecting human antibody response. V. Combined deficiencies of pantothenic acid and pyridoxine. *Am. J. Clin. Nutr.* 11:187.

Holm, G., and Palmblad, J. 1976. Acute energy deprivation in man: Effect on cell-mediated immunological reactions. *Clin. Exp. Immunol.* 25:207.

Homburger, F. 1974. "Modifiers of Carcinogenesis," Karger, Basel, p. 110.

Horwitt, M. K. 1955. Implications of observations made during experimental deficiencies in man. *Ann. N.Y. Acad. Sci.* 63:165.

Howell, D. 1971. Consequences of mild iron deficiency in children *in* "Proceedings of Workshop on Extent and Meaning of Iron Deficiency in the U.S." (G. Goldsmith, ed.), National Research Council, Washington, D.C., p. 65.

Huber, H., and Fudenberg, H. H. 1970. The interaction of monocytes and macrophages with immunoglobulins and complement. *Ser. Haematol.* 3:160.

Hughes, W. T., Price, R. A., and Kim, K. H. 1973. *Pneumocystis carinii* pneumonitis in children with malignancies. *J. Pediatr.* 82:404.

Hughes, W. T., Price, R. A., Sisko, F., Havron, W. S., Kafatos, A. G., Schonland, M., and Smythe, P. M. 1974. Protein-calorie malnutrition: A host determinant for *Pneumocystis carinii* infection. *Am. J. Dis. Child.* 128:44.

Humbert, J. R., Miescher, P. A., and Jaffe, E. R., eds., 1975. "Neutrophil Physiology and Pathology," Grune and Stratton, New York.

Jacob John, T. 1975. Oral polio vaccination of children in the tropics. II. Antibody response in relation to vaccine virus infection, *Am. J. Epidemiol.* 102:414.

Jacob John, T. 1976. Antibody responses of infants in tropics to five doses of oral polio vaccine. *Br. Med. J.* 2:812.

James, W. P. T., Drasar, B. S., and Miller, C. 1972. Physiological mechanism and pathogenesis of weanling diarrhoea. *Am. J. Clin. Nutr.* 25:564.

Janeway, C. A., Jr., Sharrow, S. O., and Simpson, E. 1975. T-cell populations with different functions. *Nature* 253:544.

Jaya, Rao, K. S. 1974. Evolution of kwashiorkor and marasmus. *Lancet* 1:709.

Jayalakshmi, V. T., and Gopalan, C. 1958. Nutrition and tuberculosis, I. An epidemiological study. *Indian J. Med. Res.* 46:87.

Jelliffe, D. B. 1966. "Assessment of Nutritional Status of the Community," Monograph Series 53, World Health Organization, Geneva.

Johannsson, S. G. O., Mellein, T., and Vahlquist, B. 1968. Immunoglobulin levels in Ethiopian preschool children with special reference to high concentrations of immunoglobulin E (IgND). *Lancet* 1:1118.

Johnston, R. B., and Stroud, R. M. 1977. Complement and host defense against infection. *J. Pediatr.* 90:169.

Jose, D. G., and Good, R. A. 1971. Absence of enhancing antibody in cell mediated immunity to tumor heterografts in protein deficient rats. *Nature* 231:323.

Jose, G. D., and Good, R. A. 1973. Quantitative effects of nutritional protein and calorie deficiency upon immune responses to tumors in mice. *Cancer Res.* 33:807.

Jose, D. G., and Welch, J. S. 1970. Growth retardation, anemia and infection, with malabsorption and infestation of the bowel. The syndrome of protein-calorie malnutrition in Australian Aboriginal children. *Med. J. Austral.* 1:349.

Jose, D. G., Welch, J. S., and Doherty, R. L. 1970. Humoral and cellular immune responses to streptococci, influenza and other antigens in Australian aboriginal school children. *Aust. Paediatr. J.* 6:192.

Jose, D. G., Stutman, O., and Good, R. A. 1973. Long term effects on immune function of early nutritional deprivation. *Nature* 241:57.

Jose, D. G., Ford, G. W., and Welch, J. S. 1976. Therapy with parent's lymphocyte transfer factor in children with infection and malnutrition. *Lancet* 1:263.

Joynson, D. H. M., Jacobs, A., Walker, D. M., and Dolby, A. E. 1972. Defect of cell-mediated immunity in patients with iron-deficiency anemia. *Lancet* 2:1058.

Jubert, A., Lee, E., and Hersh, E. 1973. Effects of surgery, anesthesia and intraoperative blood loss on immunocompetence. *J. Surg. Res.* 15:399.

Jurin, M., and Tannock, I. F. 1972. Influence of vitamin A on immunological response. *Immunology* 23:283.

Kalden, J. A., and Guthy, E. A. 1972. Prolonged skin allograft survival in vitamin C-deficiency guinea pigs. *Eur. Surg. Res.* 4:114.

Kantor, F. S. 1975. Infection, anergy and cell-mediated immunity. *N. Engl. J. Med.* 292:629.

Katz, M., Keusch, G. T., and Mata, L. J., eds. 1975. Malnutrition and infection during pregnancy: Determinants of growth and development of the child. *Am. J. Dis. Child.* 129:419.

Keet, M. P., and Thom, H. 1969. Serum immunoglobulins in kwashiorkor. *Arch. Dis. Child.* 44:600.

Kendall, A. C., and Nolan, R. 1972. Polymorphonuclear leucocyte activity in malnourished children. *Cent. Afr. J. Med.* 18:73.

Kenny, M. A., Roderbuck, C. E., Arnrich, L., and Piedad, F. 1968. Effect of protein deficiency on the spleen and antibody formation in rats. *J. Nutr.* 95:173.

Kenny, M. A., Magee, J. L., and Piedad-Pascual, F. 1970. Dietary amino acids and immune response in rats. *J. Nutr.* 100:1063.

Kent, D. C., and Schwartz, R. 1967. Active pulmonary tuberculosis with negative tuberculin skin reactions. *Am. Rev. Respir. Dis.* 95:411.

Keusch, G. T., Weinstein, L., and Grady, G. F. 1971. Biochemical effects of cholera toxin. II. Glucose metabolism in the intestine of infant rabbits. *J. Infect. Dis.* 124:188.

Keusch, G. T., Urrutia, J. J., Fernandes, R., Guerrero, O., and Casteneda, G. 1977a. Humoral and cellular aspects of intracellular bacterial killing in Guatemalan children with protein-calorie malnutrition *in* "Malnutrition and the Immune Response" (R. M. Suskind, ed.), Raven Press, New York, p. 245.

Keusch, G. T., Urrutia, J. J., Guerrero, O., Casteneda, G., and Douglas, S. D. 1977b. Rosette-forming lymphocytes in Guatemalan children with protein-calorie malnutrition *in* "Malnutrition and the Immune Response" (R. M. Suskind, ed.), Raven Press, New York, p. 117.

Kielman, A. A., Uberoi, I. S., Chandra, R. K., and Mehra, V. L. 1976. The effect of nutritional status on immune capacity and immune responses in preschool children in a rural community in India. *Bull. W.H.O.*, 54:477.

Kikkawa, Y., Kamimura, K., Hamajima, T., Sakiguchi, T., Kawai, T., Takenaka, M., and Tada, T. 1973. Thymic alymphoplasia with hyper-IgE-globulinemia. *Pediatrics* 51:690.

Kjösen, B., Bassöe, H. H., and Myking, O. 1975. The glucose oxidation in isolated leukocytes from female patients suffering from overweight or anorexia nervosa. *Scand. J. Clin. Lab. Invest.* 35:447.

Klebanoff, S. J. 1975. Antimicrobial mechanisms in neutrophilic polymorphonuclear leukocytes. *Semin. Hematol.* 12:117.

Klimentova, A. A., and Frjazinova, I. B. 1963. [The effect of dietary protein deficiency on immunogenesis and cellular changes in rat lymphoid tissue.] *Vopr. Infekts. Patol. Immunol.* p. 45.

Klimentova, A. A., and Frjazinova, I. B. 1965. [Immunogenesis and cellular reaction of the lymph nodes in C-hypovitaminosis.] *Zh. Mikrobiol. Immunobiol.* 42:96.

Klopper, A., ed., 1976. "Plasma Hormone Assays in Evaluation of Fetal Well Being," Pitman, London.

Koerner, T. A., Getz, H. R., and Long, E. R. 1949. Experimental studies on nutrition and tuberculosis. The role of protein in resistance to tuberculosis. *Proc. Soc. Exp. Biol. Med.* 71:154.

Krakower, C., Hoffman, W. A., and Axtmayer, J. H. 1940. The fate of schistosomes (*S. mansonia*) in experimental infections of normal and vitamin A deficient white rats. *Puerto Rico J. Public Health Trop. Med.* 16:269.

Krishnan, S., Bhuyan, U. N., Talwar, G. P., and Ramalingaswami, V. 1974. Effect of vitamin A and protein-calorie undernutrition on immune response. *Immunology* 27:383.

Kulapongs, P., Vithayasai, V., Suskind, R., and Olson, R. E. 1974. Cell-mediated immunity and phagocytosis and killing function in children with severe iron-deficiency anemia. *Lancet* 2:689.

Kulapongs, P., Suskind, R. M., Vithayasai, V., and Olson, R. E., 1977a. *In vitro* cell-mediated immune response in Thai children with protein-calorie malnutrition *in* "Malnutrition and the Immune Response" (R. M. Suskind), Raven Press, New York, p. 99.

Kulapongs, P., Edelman, R., Suskind, R., and Olson, R. E. 1977b. Defective

local leukocyte mobilization in children with kwashiorkor. *Am. J. Clin. Nutr.* 30:367.
Kumate, J. 1969. Desnutrición e immunología *in* "Desnutrición en el Niño" (R. Ramos-Galván, ed.), Impresiones Modernas, Mexico, p. 121.
Kutty, K. M., Chandra, R. K., and Chandra, S. 1976. Acetylcholinesterase in erythrocytes and lymphocytes: Its contribution to cell membrane structure and function. *Experientia* 32:289.
Lachmann, P. J. 1975. Complement *in* "Clinical Aspects of Immunology" (P. G. H. Gell, R. A. A. Coombs, and P. J. Lachmann, eds.), Blackwell, Oxford, p. 323.
Larsen, H. E., and Blades, R. 1976. Impairment of human polymorphonuclear leucocyte function by influenza virus. *Lancet* 1:283.
Law, D. K., Dudrick, S. J., and Abdow, N. I. 1973. Immunocompetence of patients with protein-caloric malnutrition. *Ann. Intern. Med.* 79:545.
Lawton, A. R., and Cooper, M. D. 1973. Development of immunity: Phylogeny and ontogeny *in* "Immunologic Disorders in Infants and Children" (E. R. Stiehm and V. A. Fulginiti, eds.), Saunders, Philadelphia, p. 28.
Lawton, A. R., Self, K. S., Royal, S. A., and Cooper, M. D. 1972. Ontogeny of B-lymphocytes in the human fetus. *Clin. Immunol. Immunopathol.* 1:84.
Leukskaja, Z. K. 1964. [The antibody level in vitamin A deficiency of chickens immunized with an antigen made from Ascaridia galli nematodes.] *Dokl. Akad. Nauk SSSR* 159:938.
Levene, G. M., Turk, J. L., Wright, D. J. M., and Grimble, A. G. S. 1969. Reduced lymphocyte transformation due to a plasma factor in patients with active syphilis. *Lancet* 2:246.
Lichstein, H. C., McCall, K. B., Kearney, E. B., Elvehjem, C. A., and Clark, P. F. 1946. Effect of minerals on susceptibility of Swiss mice to Theilers virus. *Proc. Soc. Exp. Biol. Med.* 62:279.
Lopez, V., Davis, S. D., and Smith, N. J. 1972. Studies in infantile marasmus. IV. Impairment of immunologic responses in the marasmic pig. *Pediatr. Res.* 6:779.
Ludovici, P. P., and Axelrod, A. E. 1951a. Circulating antibodies in vitamin-deficient states: Pteroylglutamic acid, niacin-tryptophan, vitamins B_{12}, A and D deficiencies. *Proc. Soc. Exp. Biol. Med.* 77:526.
Ludovici, P. P., and Axelrod, A. E. 1951b. Relative effectiveness of pantothenic acid and pantothol in stimulating antibody response of pantothenic deficient rats, *Proc. Soc. Exp. Biol. Med.* 77:530.
Ludovici, P. P., Axelrod, A. E., and Carter, B. B. 1949. Circulating antibodies in vitamin deficiency states. Pantothenic acid deficiency. *Proc. Soc. Exp. Biol.* 72:81.
MacCuish, A. C., Urbaniak, S. J., Goldstone, A. H., and Irvine, W. J. 1974. PHA responsiveness and subpopulations of circulating lymphocytes in pernicious anemia. *Blood.* 44:849.
MacDougall, L. G., Anderson, R., McNab, G. M., and Katz, J. 1975. Immune response in iron-deficient children: Impaired cellular defense mechanisms with altered humoral components. *J. Pediatr.* 86:833.

MacKay, H. M. 1921. The effect on kittens of a diet deficient in animal fat. *Biochem. J.* 15:19.
MacKay, H. M. 1928. Anaemia in infancy: Its prevalence and prevention. *Arch. Dis. Child.* 3:117.
Madden, S. C., and Whipple, G. H. 1940. Plasma proteins: Their source, production and utilization. *Physiol. Rev.* 20:194.
Makinodan, T., Perkins, E. H., and Chen, M. G. 1971. Immunologic activity of the aged. *Adv. Gerontol. Res.* 3:171.
Malavé, I., and Layrisse, M. 1976. Immune response in malnutrition. Differential effect of dietary protein restriction on the IgM and IgG response to alloantigens. *Cell. Immunol.* 21:337.
Mameesh, M. S., Metcoff, J., Costiloe, P., and Crosby, W. 1976. Kinetic properties of pyruvate kinase in human maternal leukocytes in human malnutrition. *Pediatr. Res.* 10:561.
Mandell, G. L., Rubin, W., and Hook, E. W. 1970. The effect of an NADH oxidase inhibitor (hydrocortisone) on polymorphonuclear leukocyte bactericidal activity. *J. Clin. Invest.* 49:1381.
Manerikar, S., Malaviya, A. N., Singh, M. B., Rajgopalan, P., and Kumar, R. 1976. Immune status and BCG vaccination in newborns with intra-uterine growth retardation. *Clin. Exp. Immunol.* 26:173.
Mangi, R. J., Niederman, J. C., Kelleher, J. E., Jr., Dwyer, J. M., Evans, A. S., and Kantor, F. S. 1974. Depression of cell-mediated immunity during acute infectious mononucleosis, *N. Engl. J. Med.* 291:1149.
Manninger, R. 1928. Zur bakteriologischen Differential-Diagnose zwischen Geflügelcholera und Hühnertyphus. *Dtsch. Tieraerztzl. Wochenschr.* 36:870.
Marks, H. H., 1960. Influence of obesity on morbidity and mortality. *Bull. N.Y. Acad. Med.* 36:296.
Marks, I. N., Brickman, H. F., and Beandry, P. H. 1974. The timing of tuberculin tests in relation to immunization with live viral vaccines. *Pediatrics* 55:392.
Masawe, A. E., Muindi, J. M., and Swai, G. B. R. 1974. Infections in iron deficiency and other types of anemia in the tropics. *Lancet* 2:314.
Mason, K. E., and Bergel, M. 1955. Maintenance of Mycobacterium leprae in rats and hamsters fed diets low in vitamin E and high in unsaturated fats. *Fed. Proc., Fed. Am. Soc. Exp. Biol.* 14:442.
Mata, L. J. 1975. Malnutrition infection interactions in the tropics. *Am. J. Trop. Med. Hyg.* 24:564.
Mata, L. J., and Faulk, W. P. 1973. The immune response of malnourished subjects with special reference to measles. *Arch. Latin-amer. Nutr.* 23:345.
Mata, L. J., Urrutia, J. J., and Lechtig, A. 1971. Infection and nutrition of children of a low socioeconomic rural community. *Am. J. Clin. Nutr.* 24:249.
Mata, L. J., Jimenez, F., and Cordon, M. 1972. Gastrointestinal flora of children with protein-calorie malnutrition. *Am. J. Clin. Nutr.* 25:118.

Mathews, J. D., Whittingham, S., Mackay, I. R., and Malcolm, L. A. 1972. Protein supplementation and enhanced antibody-producing capacity in New Guinean school-children. *Lancet* 1:675.

Mathews, J. D., Mackay, I. R., Tucker, L., and Malcolm, L. A. 1974. Interrelationship between dietary protein, immunoglobulin levels, humoral immune responses and growth in New Guinean school children. *Am. J. Clin. Nutr.* 27:908.

Mathur, M., Ramalingaswami, V., and Deo, M. G. 1972. Influence of protein deficiency on 19S antibody-forming cells in rats and mice. *J. Nutr.* 102:841.

Maugh, T. H. 1974. Vitamin A: Potential protection from carcinogens. *Science* 186:1198.

McCall, C. E., Katayama, I., Cotran, R. S., and Finland, M. 1969. Lysosomal and ultrastructural changes in human "toxic" neutrophils during bacterial infection. *J. Exp. Med.* 129:267.

McCall, C. E., Caves, J., Cooper, R., and DeChatelet, L. 1971. Functional characteristics of human toxic neutrophils. *J. Infect. Dis.* 124:68.

McCance, R. A., and Widdowson, E. M. 1964. Protein metabolism and requirements in the newborn *in* "Mammalian Protein Metabolism," Vol. II (H. Munro and J. Allison, eds.), Academic Press, New York.

McClung, L. S., and Winter, J. C. 1932. Effect of vitamin A-free diet on resistance to infections of Salmonella enteritidis. *J. Infect. Dis.* 51:475.

McFarlane, H. 1973. Immunoglobulins in populations of subtropical and tropical countries. *Adv. Clin. Chem.* 16:153.

McFarlane, H., and Hamid, J. 1973. Cell-mediated immune response in malnutrition. *Clin. Exp. Immunol.* 13:153.

McFarlane, H., Reddy, S., Adcock, K. J., Adeshina, H., Cooke, A. R., and Akene, J. 1970. Immunity, transferrin, and survival in kwashiorkor. *Br. Med. J.* 4:268.

McGregor, I. A., Rowe, D. S., Wilson, M. E., and Billewicz, W. Z. 1970. Plasma immunoglobulin concentrations in an African (Gambian) community in relation to season, malaria and other infections and pregnancy. *Clin. Exp. Immunol.* 7:51.

McGuire, E. A., Young, V. R., Newberne, P. M., and Payne, B. J. 1968. Effects of Salmonella typhimurium infection in rats fed varying protein intakes. *Arch. Pathol.* 86:60.

Meares, E. M., Jr. 1975. Factors that influence surgical wound infections. *Urology.* 6:535.

Medawar, P. B., and Simpson, E. 1975. Thymus-dependent lymphocytes. *Nature* 258:106.

Mehrishi, J. N., and Zeiller, K. 1974. Surface molecular components of T and B lymphocytes. *Eur. J. Immunol.* 4:474.

Metcoff, J. 1974. Maternal leukocyte metabolism in fetal malnutrition: Nutrition and malnutrition, identification and measurement. *Adv. Exp. Med. Biol.* 49:73.

Mettrick, D. S., and Munro, H. N. 1965. Studies on the protein metabolism of cestodes. I. Effect of host dietary constituents on the growth of *Hymenoledis diminuta*. *Parasitology* 55:453.

Meyer, J., Malgras, J., and Hirtz, G. 1955. La pantothénthérapie et son rapport avec la choinestérasémie, la protidémie et la génèse des anticorps. *Rev. Immunol.* 19:337.

Meyer, J., Malgras, J., and Ponelle, J. 1956. Action de l'acide pantothénique dans l'immunisation spécifique. *Rev. Immunol.* 20:55.

Mitchell, H. H. 1955. Some species and age differences in amino acid requirements *in* "Protein and Amino Acid Requirements of Mammals (A.A. Albanese, ed.), Academic Press, New York, p. 1.

Mohanram, M., Reddy, V., and Mishra, S. 1974. Lysozyme activity in plasma and leucocytes in malnourished children, *Br. J. Nutr.* 32:313.

Moore, D. L., Heyworth, D., and Brown, J. 1974. PHA-induced lymphocyte transformation in leucocyte cultures from malarious, malnourished and control Gambian children. *Clin. Exp. Immunol.* 17:647.

Morehead, C. D., Morehead, M., Allen, D. M., and Olson, R. E. 1974. Bacterial infections in malnourished children, *Env. Ch. Hlth.* 20:141.

Morey, G. R., and Spies, T. D. 1942. Antibody response of persons with pellagra, beriberi and riboflavin deficiency. *Proc. Soc. Exp. Biol. Med.* 49:519.

Morgan, A. G., and Soothill, J. F. 1975. Relationship between macrophage clearance of PVP and affinity of anti-protein antibody response in inbred mouse strains. *Nature* 254:711.

Morley, D. C. 1962. Measles in Nigeria. *Am. J. Dis. Child.* 103:230.

Morley, D. C. 1964. The severe measles of West Africa. *Proc. R. Soc. Med.* 57:846.

Mortensen, R. F., Osmand, A. P., and Gewurz, H. 1975. Effects of C-reactive protein on the lymphoid system. I. Binding to thymus-dependent lymphocytes and alteration of their functions. *J. Exp. Med.* 141:821.

Muller-Eberhard, H. J. 1975. Complement. *Annu. Rev. Biochem.* 44:697.

Munro, H. 1971. Impact of nutritional research on human health and survival. *Fed. Proc., Am. Soc. Exp. Biol.* 30:1403.

Murray, M. J., and Murray, A. B. 1977. Starvation suppression and refeeding activation of infection, an ecological necessity? *Lancet* 1:123.

Mutsuyama, M., Wiadrowski, M. N., and Metcalf, D. 1966. Autoradiographic analysis of lymphopoiesis and lymphocyte migration in mice bearing multiple thymus grafts. *J. Exp. Med.* 123:559.

Naeye, R. L., Diener, M. M., Harcke, H. T., Jr., and Blanc, W. A. 1971. Relation of poverty and race to birth weight and organ and cell structure in the newborn. *Pediatr. Res.* 5:17.

Naeye, R. L., Blanc, W., and Paul, C. 1973. Effects of maternal nutrition on the human fetus. *Pediatrics* 52:494.

Najjar, S. S., Stephan, M., and Asfour, R. Y. 1969. Serum levels of immunoglobulins in marasmic infants. *Arch. Dis. Child.* 44:120.

Nalder, B. N., Mahoney, A. W., Ramakrishnan, R., and Hendricks, D. G. 1972.

Sensitivity of the immune response to the nutritional status of rats. *J. Nutr.* 102:535.

Nelson, D. S., ed., 1976. "Immunobiology of the Macrophage," Academic Press, New York.

Nessan, V. J., Geerken, R. C., and Ulvilla, J. 1974. Uric acid excretion in infectious mononucleosis: A function of increased purine turnover. *J. Clin. Endocrinol. Metab.* 38:652.

Neufeld, H. A., Pace, J. A., and White, F. E. 1976. The effect of bacterial infections on ketone concentrations in rat liver and blood and on free fatty acid concentrations in rat blood. *Metab. Clin. Exp.* 25:877.

Neumann, C. G., Lawlor, G. J., Stiehm, E. R., Swedseid, M. E., Newton, C., Herbert, J., Ammann, A. J., and Jacob, M. 1975. Immunological responses in malnourished children. *Am. J. Clin. Nutr.* 28:89.

Neumann, C. G., Stiehm, E. R., and Swenseid, M. 1977. Complement levels in Ghanaian children with protein-calorie malnutrition *in* "Malnutrition and the Immune Response" (R. M. Suskind, ed.), Raven Press, New York, p. 333.

Newberne, P. M. 1966. Overnutrition and resistance of dogs to distemper virus. *Fed. Proc., Fed. Am. Soc. Exp. Biol.* 25:1701.

Newberne, P. M. 1975. Animal models for investigation of latent effects of malnutrition. *Am. J. Dis. Child.* 129:574.

Newberne, P. M. 1977. Effects of folic acid, B_{12}, choline and methionine on immunocompetence and cell-mediated immunity *in* "Malnutrition and the Immune Response" (R. M. Suskind, ed.), Raven Press, New York, p. 374.

Newberne, P. M., and Gebhardt, B. M. 1973. Pre- and post-natal malnutrition and response to infection. *Nutr. Rep. Int.* 7:407.

Newberne, P. M., and Williams, G. 1970. Nutritional influences on the course of infection *in* "Resistance to Infectious Diseases" (R. H. Dunlop and H. W. Moon, eds.), Saskatoon Modern Press, Univ. of Saskatchewan, Saskatoon.

Newberne, P. M., and Wilson, R. B. 1972. Prenatal malnutrition and postnatal responses to infection. *Nutr. Rep. Int.* 5:151.

Newberne, P. M., Hunt, C. E., and Young, V. R. 1968. The role of diet and the reticuloendothelial system in the responce of rats to Salmonella typhimurium infection. *Br. J. Exp. Pathol.* 49:448.

Newberne, P. M., Ahlstrom, A., and Rogers, A. E. 1970a. Effects of maternal dietary lipotropes on prenatal and neonatal rats. *J. Nutr.* 100:1089.

Newberne, P. M., Wilson, R. B., and Williams, G. 1970b. Effects of severe and marginal maternal lipotrope deficiency on response of postnatal rats to infection. *Br. J. Exp. Pathol.* 51:22.

Newcomb, R. H., Ishizaka, K., and DeVald, B. L. 1969. Human IgG and IgA diphtheria antitoxins in serum, nasal fluids, and saliva. *J. Immunol.* 103:215.

Nutrition Canada. 1973. "Nutrition: A National Priority," Information Canada, Ottawa.

Oberle, M. W., Graham, G. G., and Levin, J. 1974. Detection of endotoxemia with the Limulus test: Preliminary studies in severely malnourished children. *J. Pediatr.* 85:570.

Olarte, J., Cravioto, J., and Campos, B. 1956. Immunidad en el niño desnutrido. I. Produccion de antitoxina difterica. *Bol. Med. Hosp. Infant. Mex.* (Span. Ed.) 13:467.

Olusi, S. O., and McFarlane, H. 1976. Effects of early protein-calorie malnutrition on the immune response. *Pediatr. Res.* 10:707.

Orr, J. B., MacLeod, J. J. R., and Mackie, T. J. 1931. Studies on nutrition in relation to immunity. *Lancet* 1:1177.

Otto, G. F. 1965. Helminthic infections *in* "Maxcy-rosenau Preventive Medicine and Public Health," 9th ed. (B. E. Sartwell, ed.), Appleton-Century-Crofts, New York, p. 258.

Palmblad, J. 1976. Fasting (acute energy deprivation) in man: Effect on polymorphonuclear granulocyte functions, plasma iron and serum transferrin. *Scand. J. Haematol.* 17:217.

Palmblad, J. 1977. Lymphomas and dietary fat. *Lancet* 1:142.

Palmblad, J., Cantell, K., Holm, G., Norberg, R., Strander, H., and Sundblad, L. 1977a. Acute energy deprivation in man: Effect on serum immunoglobulins, antibody response, complement factors 3 and 4, acute phase reactants and interferon producing capacity of blood lymphocytes. *Clin. Exp. Immunol.*, in press.

Palmblad, J., Fohlin, L., and Lundström, M. 1977b. Anorexia nervosa and polymorphonuclear (PMN) granulocyte reactions. *Scand. J. Haematol.*, in press.

Panda, B., and Combs, G. F. 1963. Impaired antibody production in chicks fed diets low in vitamin A, pantothenic acid or riboflavin. *Proc. Soc. Exp. Biol. Med.* 113:530.

Panda, B., Holmes, G. F., and DeVolt, H. M. 1964. Studies on coccidiosis in vitamin A nutrition of broilers. *Poult. Sci.* 43:154.

Parkinson, R. S., Kern, L. B., and Bowring, A. C. 1972. Intravenous alimentation in the neonate and infant. *Med. J. Aust.* 1:1182.

Passwell, J. H., Steward, M. W., and Soothill, J. F. 1974. The effects of protein malnutrition on macrophage function and the amount and affinity of antibody response. *Clin. Exp. Immunol.* 17:491.

Payne, P. R. 1970. Protein and amino acid requirements of experimental animals *in* "Nutrition and Disease in Experimental Animals" (W. D. Tavernor, ed.), Bailliere, Tindall and Cassell, London, p. 9.

Pearce, M. L., and Dayton, S. 1971. Incidence of cancer in men on a diet high in polyunsaturated fat. *Lancet* 1:464.

Pearson, H. A., and Robinson, J. E. 1976. The role of iron in host resistance. *Adv. Pediat.* 23:1.

Phillips, I., and Wharton, B. A. 1968. Acute bacterial infection in kwashiorkor and marasmus. *Br. Med. J.* 1:407.

Pitkin, R. M. 1976. Abdominal hysterectomy in obese women. *Surg. Gynec. Obstet.* 142:532.

Pitt, J. 1977. Biology of the monocyte and macrophage: A review *in* "Malnutrition and the Immune Response" (R. M. Suskind, ed.), Raven Press, New York, p. 225.

Platt, B. S., and Heard, C. R. C. 1965. The contribution of infections to protein calorie deficiency. *Trans. R. Soc. Trop. Med. Hyg.* 59:571.
Pollard, T. D., and Weihing, R. R. 1974. Actin and myosin and cell movement, *CRC Crit. Rev. Biochem.* 2:1.
Ponder, E., and Ponder, R. O. 1943. The cytology of the polymorphonuclear leukocyte in toxic conditions. *J. Lab. Clin. Med.* 28:316.
Pretorius, P. J., and de Villiers, L. S. 1962. Antibody response in children with protein malnutrition. *Am. J. Clin. Nutr.* 10:379.
Price, P., and Bell, R. G. 1976. The effects of nutritional rehabilitation on antibody production in protein-deficient mice. *Immunology* 31:953.
Printen, K. J., Paulk, S. C., and Mason, E. E. 1975. Acute postoperative wound complications after gastric surgery for morbid obesity. *Am. Surg.* 41:483.
Pruzansky, J., and Axelrod, A. E. 1955. Antibody production to diphtheria toxoid in vitamin deficiency states. *Proc. Soc. Exp. Biol. Med.* 89:323.
Puffer, P. R., and Serrano, C. V. 1973. "Patterns of Mortality in Childhood." Pan American Health Organization, Washington D.C.
Purkaysatha, S., Kapoor, B. M. L., and Deo, M. G. 1975. Influence of protein deficiency on homograft rejection and histocompatibility antigens in rats. *Indian J. Med. Res.* 63:1150.
Purtilo, D. T., and Connor, D. H. 1975. Fatal infections in protein-calorie malnourished children with thymolymphatic atrophy. *Arch. Dis. Child.* 50:149.
Quie, P. G., Messner, R. P., and Williams, R. C. 1968. Phagocytosis in subacute bacterial endocarditis: Localization of the primary opsonic site to Fc fragment. *J. Exp. Med.* 128:553.
Raica, N., Scott, J., Lowry, D. S., and Sauberlich, H. E. 1972. Vitamin A concentration in human tissues collected from five areas in the United States. *Am. J. Clin. Nutr.* 25:291.
Ramalingaswami, V. 1964. Perspectives in protein malnutrition. *Nature* 201:546.
Rambaud, J. C., and Matuchansky, C. 1973. Alpha-chain disease: Pathogenesis and relation to Mediterranean lymphoma. *Lancet* 2:1430.
Ranki, A., Totterman, T. H., and Hayry, P. 1976. Identification of resting human T and B lymphocytes by acid α-naphthyl acetate esterase staining combined with rosette formation with *Staphylococcus aureus* strain Cowan 1. *Scand. J. Immunol.* 5:1129.
Rao, B. S. N., and Gopalan, C. 1958. Nutrition and tuberculosis. II. Studies on nitrogen, calcium and phosphorus metabolism in tuberculosis. *Indian J. Med. Res.* 46:93.
Rao, K. S. J., Srikantia, S. G., and Gopalan, C. 1968. Plasma cortisol levels in protein-calorie malnutrition. *Arch. Dis. Child.* 43:365.
Rapoport, M. I., and Beisel, W. R. 1971. Studies on tryptophan metabolism in experimental animals and man during infectious illness. *Am. J. Clin. Nutr.* 24:807.
Rapoport, M. I., Lust, G., and Beisel, W. R. 1968. Host enzyme induction of bacterial infection. *Arch. Intern. Med.* 121:11.
Ratcliffe, H. L., and Merrick, J. V. 1957. Tuberculosis induced by droplet nuclei

infection. Its development pattern in hamsters in relation to levels of dietary protein. *Am. J. Pathol.* 33:107.

Ratnaker, K. S., Mathur, M., Ramalingaswami, V., and Deo, M. G. 1972. Phagocytic function of reticuloendothelial system in protein deficiency: A study in rhesus monkeys using ^{32}P-labeled *E. coli. J. Nutr.* 102:1233.

Rayfield, E. J., Curnow, R. T., George, D. T., and Beisel, W. R. 1973. Impaired carbohydrate metabolism during a mild viral illness. *N. Engl. J. Med.* 289:618.

Reddy, V., and Srikantia, S. G. 1964. Antibody response in kwashiorkor. *Indian J. Med. Res.* 52:1154.

Ritterson, A. L., and Stauber, L. A. 1949. Protein intake and leishmaniasis in the hamster. *Proc. Soc. Exp. Biol. Med.* 70:47.

Robertson, E. C., and Ross, J. R. 1932. The effect of vitamin D in increasing resistance to infection. *J. Pediatr.* 1:69.

Robson, L. C., and Schwarz, M. R. 1975a. Vitamin B_6 deficiency and the lymphoid system. I. Effects on cellular immunity and *in vitro* incorporation of ^{3}H-uridine by small lymphocytes. *Cell. Immunol.* 16:145.

Robson, L. C., and Schwarz, M. R. 1975b. Vitamin B_6 deficiency and the lymphoid system. II. Effects of vitamin B_6 deficiency *in utero* on the immunological competence of the offspring. *Cell. Immunol.* 16:145.

Rocha, D. M., Santeusanio, F., Faloona, G. R., and Unger, R. H. 1973. Abnormal pancreatic alpha-cell function in bacterial infections. *N. Engl. J. Med.* 288:700.

Roos, A., Hegsted, D. M., and Stare, F. J. 1946. Nutritional studies with the duck. IV. The effect of vitamin deficiencies on the course of *P. lophurae* infection in the duck and the chick. *J. Nutr.* 32:473.

Rose, G., Blackburn, H., Keys, A., Taylor, H. L., Kannel, W. B., Paul, O., Reid, D. D., and Stamler, J. 1974. Colon cancer and blood cholesterol. *Lancet* 1:181.

Rosen, E. U., Geefhuysen, J., and Ipp, I. 1971. Immunoglobulin levels in protein calorie malnutrition. *S. Afr. Med. J.* 45:980.

Rosen, E. U., Geefhuysen, J., Anderson, R., Joffe, M., and Rabson, A. R. 1975. Leucocyte function in children with kwashiorkor. *Arch. Dis. Child.* 50:220.

Ross, M. H., and Bras, G. 1971. Lasting influence of early caloric restriction on prevalence of neoplasm in the rat. *J. Nat. Cancer Inst.* 47:1095.

Ross, M. H., and Bras, G. 1973. Influence of protein under- and over-nutrition on spontaneous tumor prevalence in the rat. *J. Nutr.* 103:944.

Rowe, D. S., McGregor, I. A., Smith, S. J., Hall, P., and Williams, K. 1968. Plasma immunoglobulin concentrations in a West African (Gambian) community and in a group of healthy British adults. *Clin. Exp. Immunol.* 3:63.

Ruckman, I. 1946. The effect of nutritional deficiencies on the development of neutralizing antibodies and associated changes in cerebral resistance against the virus of Western equine encephalomyelitis. *J. Immunol.* 53:51.

Ruebner, B. H., and Miyai, K. 1961. Effect of amino acids on growth and susceptibility to viral hepatitis in mice. *J. Lab. Clin. Med.* 58:627.

Sabin, A. B. 1959. Present position of immunization against poliomyelitis with live poliovirus vaccine. *Br. Med. J.* 1:663.
Sabin, A. B., and Duffy, C. E. 1940. Nutrition as a factor in the development of constitutional barriers to involvement of the nervous system by certain viruses. *Science* 91:552.
Schade, A. K., and Caroline, L. 1946. An iron-binding component in human blood plasma. *Science* 104:340.
Schaedler, R. W., and DuBois, R. J. 1959. Effect of dietary proteins and amino acids on the susceptibility of mice to bacterial infections. *J. Exp. Med.* 110:921.
Schlesinger, L., and Stekel, A. 1974. Impaired cellular immunity in marasmic infants. *Am. J. Clin. Nutr.* 27:615.
Schlesinger, L., Ohlbaum, A., Grez, L., and Stekel, A. 1977. Cell-mediated immune studies in marasmic children from Chile: Delayed hypersensitivity, lymphocyte transformation and interferon production *in* "Malnutrition and the Immune Responses (R. M. Suskind, ed.), Raven Press, New York, p. 91.
Schneider, R. E., and Viteri, F. E. 1974. Luminal events of lipid absorption in protein-calorie malnourished children. Relationship with nutritional recovery and diarrhoea. *Am. J. Clin. Nutr.* 27:777.
Schonland, M., Shanley, B. C., Loening, W. E. K., Parent, M. A., and Coovadia, H. M. 1972. Plasma cortisol and immunosuppression in protein-calorie malnutrition. *Lancet* 2:435.
Schopfer, K., and Douglas, S. D. 1975. Immunological aspects of infantile protein-calorie malnutrition. *Bull. Schweiz. Akad. Med. Wiss.* 31:327.
Schopfer, K., and Douglas, S. D. 1976a. *In vitro* studies of lymphocytes from children with kwashiorkor. *Clin. Immunol. Immunopathol.* 5:21.
Schopfer, K., and Douglas, S. D. 1976b. Fine structural studies of peripheral blood leucocytes from children with kwashiorkor: Morphological and functional properties. *Br. J. Haematol.* 32:573.
Schopfer, K., and Douglas, S. D. 1976c. Neutrophil function in children with kwashiorkor. *J. Lab. Clin. Med.* 88:450.
Scrimshaw, N. S. 1970. Synergism of malnutrition and infection: Evidence from field studies in Guatemala. *J. Am. Med. Assoc.* 212:1685.
Scrimshaw, N. S., Taylor, C. E., and Gordon, J. E. 1968. "Interactions of Nutrition and Infection," Monograph Series 57, World Health Organization, Geneva.
Sellmeyer, E., Bhettay, E., Truswell, A. S., Meyers, O. L., and Hansen, J. D. L. 1972. Lymphocyte transformation in malnourished children, *Arch. Dis. Child.* 47:429.
Selvaraj, R. J., and Bhat, K. S. 1972a. Metabolic and bacterial activities of leukocytes in protein-calorie malnutrition. *Am. J. Clin. Nutr.* 25:166.
Selvaraj, R. J., and Bhat, K. S. 1972b. Phagocytosis and leukocyte enzymes in protein-calorie malnutrition. *Biochem. J.* 127:255.
Selvaraj, R. J., and Sabarra, A. J. 1966. Relationship of glycolytic and oxidative

metabolism to particle entry and destruction in phagocytizing cells. *Nature* 211:1272.

Seth, V., and Chandra, R. K. 1972. Opsonic activity, phagocytosis and intracellular bactericidal capacity of polymorphs in undernutrition. *Arch. Dis. Child.* 47:282.

Sharp, G. W. 1973. Action of cholera toxin on fluid and electrolyte movement in the small intestine. *Annu. Rev. Med.* 24:19.

Sherman, H. C., and Burtis, M. P. 1927-28. Vitamin A in relation to growth and to subsequent susceptibility to infection. *Proc. Soc. Exp. Biol. Med.* 25:649.

Shils, M. E. 1973. "Nutrition and Neoplasia," Lea and Febiger, Philadelphia, p. 981.

Shousha, S., and Kamel, K. 1972. Nitro blue tetrazolium test in children with kwashiorkor with a comment on the use of latex particles in the test. *J. Clin. Pathol.* 25:494.

Siegel, H., Squibb, R. L., Solotorovsky, M., and Ott, W. H. 1968. A quantitative pathologic study of avian tuberculosis in the chick. *Am. J. Pathol.* 52:349-367.

Sigel, M. M., and Good, R. A. eds. 1972. "Tolerance, Autoimmunity and Aging," Charles C Thomas, Springfield, Ill.

Sinha, D. P., and Bang, F. B. 1976. Protein and caloric malnutrition, cell-mediated immunity, and B.C.G. vaccination in children from rural West Bengal. *Lancet* 2:531.

Sirisinha, S., Suskind, R., Edelman, R., Charupatana, C., and Olson, R. E. 1973. Complement and C3 proactivator levels in children with protein-calorie malnutrition and effect of dietary treatment. *Lancet* 1:1016.

Sirisinha, S., Suskind, R., Edelman, R., Asvapaka, C., and Olson, R. E. 1975. Secretory and serum IgA in children with protein-calorie malnutrition. *Pediatrics* 55:166.

Smith, N. J., Khadroui, S., Lopez, V., and Hamza, B. 1977. Cellular immune response in Tunisian children with severe infantile malnutrition *in* "Malnutrition and the Immune Response" (R. M. Suskind, ed.), Raven Press, New York, p. 105.

Smythe, P. M., and Campbell, J. A. 1959. The significance of the bacteremia of kwashiorkor. *S. Afr. Med. J.* 33:777.

Smythe, P. M., Breton-Stiles, G. G., Grace, H. J., Mafoyane, A., Schonland, M., Coovadia, H. M., Loening, W. E. K., Parent, M. A., and Vos, G. H. 1971. Thymolymphatic deficiency and depression of cell-mediated immunity in protein-calorie malnutrition. *Lancet* 2:939.

Solberg, C. O., and Hellum, K. B. 1972. Neutrophil granulocyte function in bacterial infections. *Lancet* 2:727.

Solotorovsky, M., Squibb, R. L., Wogan, G. N., Seigel, H., and Gala, R. 1961. The effect of dietary fat and vitamin A on avian tuberculosis in chicks. *Am. Rev. Respir. Dis.* 84:226.

Soothill, J. F. 1975. Immunity deficiency states *in* "Clinical Aspects of Immunology" (P. G. H. Gell, R. R. A. Coombs, and P. J. Lachmann, eds.), Blackwell, Oxford, p. 649.

Squibb, R. L. 1964. Nutrition and biochemistry of survival during New Castle disease virus infection. III. Relation of dietary protein to nucleic acid and free amino acids of avian liver. *J. Nutr.* 82:442.

Sriramachari, S., and Gopalan, C. 1958. Nutrition and tuberculosis. III. Effect of some nutritional factors on resistance to tuberculosis. *Indian J. Med. Res.* 46:105.

Srivastava, U., Spach, C., and Aschkenasy, A. 1975. Cyclic AMP metabolism and nucleic acid content in the lymphocytes of the thymus, spleen, and lymph nodes of protein deficient rats. *J. Nutr.* 105:924.

Stavitsky, A. B. 1957. Participation of popliteal lymph node and spleen in production of diphtheria antitoxin in the rabbit. *J. Infect. Dis.* 94:306.

Stiehm, E. R., and Fulginiti, V. A. eds., 1973. "Immunologic Disorders in Infants and Children," Saunders, Philadelphia.

Stoerk, H. C., and Eisen, H. N. 1946. Suppression of circulating antibodies in pyridoxine deficiency. *Proc. Soc. Exp. Biol. Med.* 62:88.

Stoerk, H. C., Eisen, H. N., and John, H. M. 1947. Impairment of antibody response in pyridoxine-deficient rats. *J. Exp. Med.* 85:365.

Stossel, T. P. 1974. Phagocytosis. *N. Engl. J. Med.* 290:717.

Stossel, T. P. 1975. Phagocytosis: recognition and ingestion, *Semin. Hematol.* 12:83.

Suda, A. K., Mathur, M., Deo, K., and Deo, M. G. 1976. Kinetics of mobilization of neutrophils and their marrow pool in protein-calorie deficiency. *Blood* 48:865.

Suskind, R., ed., 1977. "Malnutrition and the Immune Response," Raven Press, New York.

Suskind, R., Olson, L. C., and Olson, R. E. 1973. Protein calorie malnutrition and infection with hepatitis-associated antigen. *Pediatrics* 54:526.

Suskind, R. M., Sirisinha, S., Vithayasai, V., Edelman, R., Damrongsak, D., Charupatana, C., and Olson, R. E. 1976a. Immunoglobulins and antibody response in children with protein-calorie malnutrition. *Am. J. Clin. Nutr.* 29:836.

Suskind, R., Edelman, R., Kulapongs, P., Pariyananda, A., and Sirisinha, S. 1976b. Complement activity in children with protein-calorie malnutrition. *Am. J. Clin. Nutr.* 29:1089.

Taneja, P. N., Ghai, O. P., and Bhakoo, O. N. 1962. Importance of measles to India. *Am. J. Dis. Child.* 103:226.

Tannenbaum, A. 1959. "Nutrition and Cancer," Hoeber-Harper, New York, p. 517.

Tannenbaum, A. 1975. "Nutrition and Cancer," Hoeber-Harper, New York.

Tannenbaum, A., and Silverstone, H. 1957. "Nutrition and the Genesis of Tumours," Butterworth, London.

Tashmukhamedov, R. R. 1965. [An experimental study of the effect produced by vitamin B_{12} on the immunogenesis in conditions of tetanus toxoid immunization.] *Zh. Mikrobiol. Immunobiol.* 42:37.

Tejada, C., Argueta, V., Sanchez, M., and Albertazzi, C. 1964. Phagocytic and alkaline phosphatase activity of leukocytes in kwashiorkor. *J. Pediatr.* 64:753.

Templeton, A. C. 1970. Generalized herpes simplex in malnourished children. *J. Clin. Pathol.* 23:24.

Thakur, M. L., Coleman, R. E., Mayhall, C. G., and Welch, M. J. 1976. Preparation and evaluation of ^{111}In-labelled leukocytes as an abscess imaging agent in dogs. *Radiology* 119:731.

Thompson, C., and Dwyer, J. M. 1975. Unpublished observations cited by Kantor (1975).

Tracey, V. V., Dew, C., and Harper, J. R. 1971. Obesity and respiratory infection in infants and young children, *Br. Med. J.* 1:16.

Trakatellis, A. C., Stinebring, W. R., and Axelrod, A. E. 1963. Studies on systemic reactivity to purified protein derivative (PPD) and endotoxin. I. Systemic reactivity to PPD in pyridoxine-deficient guinea pigs. *J. Immunol.* 91:39.

Triger, D. R., and Wright, R. 1973. Hyperglobulinemia in liver disease. *Lancet* 1:1494.

Turner, M. W., and Hulme, B. 1971. "The Plasma Proteins," Pitman, London, p. 71.

Turner, R. G., Anderson, D. E., and Loew, E. R. 1930. Bacteria of the upper respiratory tract and middle ear of albino rats deprived of vitamin A. *J. Infect. Dis.* 46:228.

Uhr, J. W., Weismann, G., and Thomas, L. 1963. Acute hypervitaminosis A in guinea pigs. II. Effects on delayed-type hypersensitivity. *Proc. Soc. Exp. Biol.* 112:287.

Underdahl, N. R., and Young, G. A. 1956. Effect of dietary intake of fat-soluble vitamins on intensity of experimental swine influenza virus infection in mice. *Virology* 2:415.

van Furth, R., ed., 1975. "Mononuclear Phagocytes in Immunity, Infection and Pathology," Blackwell, Oxford.

van Furth, R., Schuit, H. R. E., and Hijmans, W. 1965. The immunological development of the human fetus. *J. Exp. Med.* 122:1173.

Vint, F. W. 1937. Postmortem findings in the natives of Kenya. *East Afr. Med. J.* 13:332.

von Pirquet, C. 1908. Das Verhalten der kutanen Tuberkulinreaktion während der Masern. *Dtsch. Med. Wochenschr.* 34:1297.

Waddell, C. C., Taunton, O. D., and Twomey, J. J. 1976. Inhibition of lymphoproliferation by hyperlipoproteinemic plasma. *J. Clin. Invest.* 58:950.

Walford, R. L. 1969. "The Immunologic Theory of Aging," Munksgaard, Copenhagen.

Walford, R. L. 1970. Antibody diversity, histocompatibility systems, disease states, and aging. *Lancet* 2:1226.

Walker, A. M., Garcia, R., Pate, P., Mata, L. J., and David, J. R. 1975. Transfer factor in the immune deficiency of protein-calorie malnutrition: A controlled study of 32 cases. *Cell. Immunol.* 15:372.

Wannemacher, R. W., Jr. 1977. Key role of various individual amino acids in host response to infection. *Am. J. Clin. Nutr.* 30:1269.

Wannemacher, R. W., Jr. Dinterman, R. E., Pekarek, R. S., and Beisel, W. R. 1974. Urinary free amino acid excretion as a measure of alterations in protein metabolism during experimentally induced infections in man. *Fed. Proc., Fed. Am. Soc. Exp. Biol.* 33:669.
Ward, P. A. 1974. Leukotaxis and leukotactic disorders, *Am. J. Path.* 77:520.
Warren, M. P., and Wiele, R. L. 1973. Clinical and metabolic features of anorexia nervosa, *Am. J. Obstet. Gynecol.* 117:435.
Waterlow, J. C., and Rutishauser, I. H. E. 1974. Malnutrition in man *in* "Early Malnutrition and Mental Development" (J. Cravioto, L. Hambraeus, and B. Vahlquist, eds.), Almqvist & Wiksell, Stockholm, p. 13.
Watts, T. 1969. Thymus weights in malnourished children. *J. Trop. Pediatr.* 15:155.
Wawszkiewicz, E. J., Schneider, H. A., Starcher, B., Pollack, J., and Neilands, J. B. 1971. Salmonellosis pacifarin activity to enterobactin. *Proc. Nat. Acad. Sci., U.S.A.* 68:2870.
Weaver, H. M. 1946. Resistance of cotton rats to the virus of poliomyelitis as affected by intake of vitamin A: Partial inanition and sex. *J. Pediatr.* 28:14.
Webb, G. B. 1916. Immunity in tuberculosis. *J. Lab. Clin. Med.* 1:414.
Weinberg, E. D. 1971. Roles of iron in host–parasite interaction. *J. Infect. Dis.* 124:401.
Werkman, C. H. 1923. Immunologic significance of vitamins. *J. Infect. Dis.* 32:247.
Werkman, C. H., Nelson, V. E., and Fulmer, E. I. 1924a. Immunologic significance of vitamins. IV. Influence of the lack of vitamin C on resistance of the guinea pig to bacterial infection, on production of specific agglutinins and on opsonic activity. *J. Infect. Dis.* 34:447.
Werkman, C. H., Baldwin, F. M., and Nelson, V. E. 1924b. Immunologic significance of vitamins. V. Resistance of the avitaminic albino rat to diphtheria toxin; production of antitoxin and blood pressure effects. *J. Infect. Dis.* 35:549.
Wertman, K., and Sarandria, J. L. 1951. Complement-fixing Murine typhus antibodies in vitamin deficiency states. *Proc. Soc. Exp. Biol. Med.* 76:388.
Widdowson, E. M. 1970. Experimental animals in the study of human nutrition *in* "Nutrition and Disease in Experimental Animals" (W. D. Tavernor, ed.), Bailliere, Tindall and Cassell, London.
Wilkinson, P. C. 1974a. Surface and cell membrane activities of leukocyte chemotactic factors. *Nature* 251:58.
Wilkinson, P. C. 1974b. "Chemotaxis and Inflammation," Churchill-Livingstone, Edinburgh.
Williams, E. A. G., Gross, R. L., and Newberne, P. M. 1975. Effect of folate deficiency on the cell-mediated immune response in rats. *Nutr. Rep. Int.* 12:137.
Williams, G. D., Newberne, P. M., and Wilson, R. B. 1972. Amino acids and carbohydrate metabolism in Salmonella infection in dogs. *Fed. Proc., Fed. Am. Soc. Exp. Biol.* 31:657.
Wissler, R. W. 1947. The effects of protein depletion and subsequent immuniza-

tion upon the response of animals to pneumococcal infection. I. Experiments with rabbits. *J. Infect. Dis.* 80:250.

Wissler, R. W., Woolridge, R. L., and Steffee, C. H. 1946. Influence of amino acid feeding upon antibody production in protein depleted rats. *Proc. Soc. Exp. Biol. Med.* 62:199.

Wissler, R. W., Fitch, F. W., LaVia, M. F., and Gunderson, C. H. 1957. The cellular basis for antibody production. *J. Cell Comp. Physiol.* 50:265.

Wohl, M. G., Reinhold. J. G., and Rose, S. B. 1949. Antibody response in patients with hypoproteinemia with special reference to the effect of supplementation with protein or protein hydrolysate. *Arch. Intern. Med.* 83:402.

Woldsdorf, J., and Nolan, R. 1974. Leucocyte function in protein deficiency states. *S. Afr. Med. J.* 48:528.

Woodruff, J. F. 1970. The influence of quantitated post-weaning undernutrition in coxsackivirus B_3 infection of adult mice. II. Alteration of host defense mechanisms. *J. Infect. Dis.* 121:164.

Woodruff, J. F., and Kilbourne, E. D. 1970. The influence of quantitated post-weaning undernutrition in coxsackvirus B_3 infection of adult mice. I. Viral persistence and increased severity of lesions. *J. Infect. Dis.* 121:137.

Woods, J. W., Welt, L. G., and Hollander, W. 1961. Susceptibility of rats to experimental pyelonephritis during potassium depletion. *J. Clin. Invest.* 40:599.

World Health Organization. 1972a. A provisional protocol for the field study of some of the effects of malnutrition on the immune response. Document IMM/72.1. Geneva.

World Health Organization. 1972b. A survey of nutritional–immunological interactions. *Bull. W.H.O.* 46:537.

World Health Organization. 1973. Food and nutrition terminology. Terminology Circular No. 27, World Health Organization, Geneva, p. 1.

Wright, W. H. 1935. The relation of vitamin A deficiency to ascariasis in the dog. *J. Parasitol.* 21:433.

Wynder, E. L., and Reddy, B. S. 1975. Dietary fat and colon cancer. *J. Nat. Cancer Inst.* 54:7.

Yaeger, R. G., and Miller, O. M. 1963. Effect of malnutrition on susceptibility of rats to Trypanosoma cruzi. V. Vitamin A deficiency. *Exp. Parasitol.* 14:9.

Yeung, C. Y. 1970. Hypoglycemia in neonatal sepsis. *J. Pediatr.* 77:812.

Yoshida, T., Metcoff, J., Frenk, S., and De la Pena, C. 1967. Intermediary metabolities and adenine nucleotides in leukocytes of children with protein-calorie malnutrition. *Nature* 214:525.

Yoshida, T., Metcoff, J., and Frenk, S. 1968. Reduced pyruvic kinase activity, altered growth patterns of ATP in leukocytes and protein-calorie malnutrition. *Am. J. Clin. Nutr.* 21:162.

Young, V. R., Chan, S. C., and Newberne, P. M. 1968. Effect of infection on skeletal muscle ribosomes in rats fed adequate or low protein diets. *J. Nutr.* 94:361.

REFERENCES

Zee, P., Walters, T., and Mitchell, C. 1970. Nutrition and poverty in preschool children. A nutritional survey of preschool children from impoverished Black families, Memphis. *J. Am. Med. Assoc.* 213:739.

Ziegler, H. D., and Ziegler, P. B. 1975. Depression of tuberculin reaction in mild and moderate protein-calorie malnourished children following BCG vaccination. *Johns Hopkins Med. J.* 137:59.

Zucker, T. K., Zucker, L. M., and Seronde, J., Jr. 1956. Antibody formation and natural resistance in nutritional deficiencies. *J. Nutr.* 59:299.

INDEX

Acetylcholinesterase, 15
Actin, 23, 50, 201
Acute-phase reactant glycoproteins, 3, 50, 54, 200
Adenosine triphosphate, 39, 125, 183
Adenyl cyclase, 39, 54, 55
Adrenaline, effect on leukocyte mobilization, 112
Agglutinins, 96
Aging, effects of
 on delayed hypersensitivity, 186
 on immunocompetence, 185
 on T lymphocyte number, 186
Albumin, 37, 48, 89
Alkaline phosphatase, heat-stable, 38
Allergy, 29, 61, 190, 201
Alpha chain disease, 189
Alloantigens, antibody response to, 98
Allograft, 8
Alpha$_1$-antitrypsin, 51
Alpha-fetoprotein, 88
Amino acids, 37, 49, 50, 54, 126, 133
Anaphylactic response, 12
Anatomic barriers, 13, 61
Anemia, 2, 5
 iron deficiency type, 7, 37, 77, 85, 86
 megaloblastic, 37, 85, 86, 160
Anergy, 57, 59, 61, 84
Anesthesia, 29, 182
Animal data, 127
Anorexia nervosa, 8, 119, 182
Antagonism, of malnutrition and infection, 2, 8, 131, 132
Anthrax, antibody response to, 97
Anthropometry, 34, 36

Antibody(ies)
 affinity of, 141
 antitetanus, 94
 blocking, 145
 to food antigens, 103, 190
 host defense and, 13
 production of, 13, 16
 properties of, 18
 response
  anemia and, 101
  iron deficiency and, 101
  nutritional deficiency and, 96, 97
  variables controlling, 96
Anticomplementary activity, 107
Antigen(s)
 on cell surface, 15
 T cell-dependent, antibody response to, 96
Antiglobulin test, 105
Antimicrobial action
 of macrophage, 25
 of neutrophils, 22, 24, 64, 116
Appendix, lymphoid tissue in, 74, 101
Apoferritin, 54
Arthus reaction, 181
Ascaris
 anemia due to, 53
 antibody response to, 98
 effect of, immunoglobulin levels and, 62, 91, 92
 infection with, 44
Ascorbic acid, deficiency of, 132, 170
Autoimmunity, 9, 61, 104, 190

Bacillus Calmette-Guerin (B.C.G.) vaccine, 84, 192

235

Bacterial infection, 44, 52, 55, 64, 181
Bacterial killing, 22, 24, 25, 64, 116
Bacteriophage ϕX174, antibody response to, 98
Basophil, 21
Bile salt metabolism
  infection and, 55
  malnutrition and, 46
Biological implications, 181
Blast transformation, 57, 58, 75, 85, 160
Blastomycosis, 57
Blood urea, 37
Body constituents, wastage of, infection and, 49
Bone marrow, 12, 16, 22, 65, 112
Breast milk, 123, 184, 191
Bronchopneumonia, 6, 42, 43
Brucella
  antibody response to, 97
  infection with, 8, 57
Burns, 29, 110
Bursa of Fabricius, 12, 13, 16, 154

C-reactive protein, 51, 126
Calcium, deficiency of, 176
Calorie deficiency, effect of
  on antibody formation, 171
  on cell-mediated immunity, 171
  on lymphoid tissues, 171
  See also Energy-protein undernutrition; Marasmus
Cancer, 9, 29, 187
  See also Tumor
Candida albicans
  delayed hypersensitivity response to, 82, 85
  infection due to, 3, 42, 44, 181
Carbohydrate intake, parasitic disease and, 152
Carbohydrate metabolism, alteration of, infection and, 51
Carbon dioxide, production of, phagocytosis and, 24, 118
Cell division, 74
Cell-mediated immunity (CMI)
  assessment of, 201
  components of, 13
  depression of, 43
  development of, 12
  in energy-protein undernutrition, 74, 80, 85

Cell-mediated immunity (CMI) (cont'd)
  evaluation of, difficulties in, 81
  in fetal malnutrition, 76, 77
  folate deficiency and, 85, 86
  infection and, 57
  in iron deficiency, 77, 85, 87
  in kwashiorkor, 81, 82
  T lymphocytes and, 12
  in marasmus, 74, 80, 85
  research needs and, 201
  response to
    Candida, 82, 85
    dinitrochlorobenzene, 75, 82, 85
    dinitrofluorobenzene, 83
    phytohemagglutinin, 57, 58, 75, 82, 85
    purified protein derivative, 81, 84
    streptococcal antigens, 82, 86
    trichophyton, 81
    tuberculin, 81, 83, 84, 145
  transferrin levels and, 83
  vitamin A deficiency and, 145
  See also Delayed hypersensitivity, T lymphocytes
Chalones, 3, 88, 200
Chediak-Higashi syndrome, 110
Chemotaxis
  complement components and, 27
  in energy-protein malnutrition, 112
  infection and, 65, 113
  in kwashiorkor, 112
  macrophage, 24, 113
  polymorphonuclear leukocyte, 21, 23, 65, 112
  serum factors and, 25
Chicken pox, 60
Children
  assessment of nutrition of, 3
  causes of mortality among, 1
  growth of, 6
  host defenses, 11
  nutritional deficiencies in, 1
Cholera toxin, 55
Choline, 162, 166
Chronic granulomatous disease, 110, 116
Cobalt, 176
Coccidiodomycosis, 57, 131
Complement, 13
  activation of, 25, 105, 108
  alternate pathway of, 27
  breakdown of, 108, 109

Complement (cont'd)
C1, 26, 27, 105, 107
C1 esterase inhibitor, 26, 27
C2, 26, 27
C3, 26, 28, 105, 107
C3 proactivator, 105, 107
C4, 26, 27, 106
C5, 26, 27, 105, 107
C6, 26, 27, 105, 107
C7, 26, 27
C8, 26, 27, 105, 107
C9, 26, 27, 105, 107
components of, 26, 27
consumption of, 108, 109
deficiency of, 104
in energy-protein undernutrition, 105, 107
functions of, 25, 27
hemolytic, 104, 105
immunoconglutinin levels and, 108, 109
infection and, 104, 105, 108
inhibitors of, 106
in kwashiorkor, 105
in marasmus, 105
synthesis of, 26, 108
Coombs test, 105, 144, 145, 190
Copper, 52, 136, 176
Cord blood, 22, 93, 94
Cornybacterium kutscheri, antibody response to, 97
Corticosteroids
  effect on
    delayed hypersensitivity, 3, 60
    immune response, 60, 126
    interferon, 3
    susceptibility to infection, 149
  infection and, 51, 54
  uptake by thymolymphatic system, 74
Cutaneous hypersensitivity, see Delayed hypersensitivity
Cystic fibrosis, 61
Cytotoxic lymphocytes, 13, 80, 161
Cytotoxicity, 12, 80, 144, 160

Degranulation, 22
Delayed hypersensitivity
  in energy-protein undernutrition, 80
  evaluation of, 81
  in iron deficiency, 85
  in kwashiorkor, 81

Delayed hypersensitivity (cont'd)
  T lymphocytes and, 15
  persistent depression of, 77, 83
  protein deficiency and, 81
  response to
    Candida antigen, 82, 85
    dinitrochlorobenzene, 75, 82, 85
    dinitrofluorobenzene, 83
    keyhole limpet hemocyanin, 82
    mumps, 81
    phytohemagglutinin, 82
    purified protein derivative, 81, 84
    streptococcal antigens, 82
    streptokinase-streptodornase, 82
    trichophyton, 81
    tuberculin, 81, 83, 84, 145
    transferrin levels and, 81
    vitamin A deficiency and, 84
Deoxyribonucleic acid (DNA), synthesis of, 59, 75, 85
Diarrhea
  in calves, 127
  in infants and children, 5, 7, 42, 46, 55, 125
  weanling, 46, 125
Dietary needs
  age and, 130
  related to growth rate, 128
  species variations in, 128
DiGeorge syndrome, 14
Dinitrochlorobenzene (DNCB), response to, 75, 82, 85
Dinitrofluorobenzene (DNFB), response to, 83
Diphtheria toxoid, antibody response to, 97
Distemper, 150, 151, 177
Döhle bodies, 64, 110, 111
Down's syndrome, 43
Dysentery, 42

Economic status, nutrition and, 2
Edema, 32, 48
Eimeria stiedie, 131
Endocrinal factors, 3, 51, 53, 54, 60, 74, 126, 149
Endogenous pyrogen, 54
Endoplasmic reticulum, 110, 111
Endotoxin
  effect on complement components, 106
  generation of chemotactic activity by, 66

Endotoxin *(cont'd)*
 infection and, 65
 mobilization of granulocytes by, 65
Energy-protein undernutrition
 antibody response in, 97
 causes of, 31
 cell-mediated immunity in, 74
 complement system in, 105, 106, 108
 definition of, 31
 delayed hypersensitivity in, 80
 diagnosis of, 32, 34
 immunoglobulin levels in, 97
 incidence of, 1
 lymph node in, 71, 72
 lymphocyte proliferation in, 85
 B lymphocytes in, 78
 T lymphocytes in, 75, 76, 78
 metabolism of immunoglobulins in, 91
 morphology of neutrophils in, 111
 phagocyte function in, 111
 secretory antibody response in, 101
 spleen in, 71, 73
 thymus in, 68, 70
 tonsil in, 71, 73
Entameba histolytica, 42
Enzymes, 23, 25, 27, 37, 50, 51, 66, 114, 115
Eosinophil, 21
Epinephrine, effect on neutrophil mobilization, 112
Epstein-Barr virus, 13, 59
Erythrocytes
 complement-coated, 17, 190
 heterologous, antibody response to, 98, 144, 163, 170
 IgG-sensitized, 17, 190
 sheep, 14, 15, 17, 144, 145, 163, 170
Escherichia coli
 antibody response to, 97
 growth of, 123
 phagocytosis of, 145
Estrogens, 143

Factor B, 27
Factor D, 27
Fat, intake of susceptibility to infection and, 150, 152
Fat, metabolism of, infection and, 52
Ferritin, 54, 55
Fetal growth retardation, *see* Fetal malnutrition

Fetal malnutrition
 antibody response in, 99, 100
 delayed hypersensitivity in, 77, 84
 diagnosis of, 38, 40
 food antibodies in, 191
 immunoglobulin levels in, 94, 95
 intergenerational effects of, 184
 T lymphocytes in, 76
 postnatal growth in, 77
 research needs and, 200
 thymus in, 70
Fetus, maternal malnutrition and, 39
Fever, metabolic changes in, 49, 53
Fibroblast, 124
Flagellin, antibody response to, 97
Folic acid, deficiency of
 cell-mediated immunity in, 86, 160
 cytotoxicity in, 160
 diagnosis of, 37
 infection and, 53
 lymphocyte proliferation in, 86, 160
 susceptibility to infection in, 160
Food antigens
 absorption of, 191
 antibodies to, 103, 190
Foot-and-mouth disease, 7
Formazon, 115, 116
Fructokinase, 114
Fructose-1,6-diphosphate, 39

Gastroenteritis, 5, 7, 42, 46, 125
Gastrointestinal microflora, 46, 125
Giardia lamblia, 42
γ-Globulins
 catabolic rate of, 89
 levels of, 89
 loss of, 90
 nutritional deficiency and, 89
 production of, 61, 89
Glucagon, 51, 54
Glucocorticoids, *see* Corticosteroids
Gluconeogenesis, infection and, 49, 51, 52
Glucose, 51
Glucose-6-phosphate dehydrogenase, 24, 65, 110
Glycine-rich-β-glucoprotein, 27
Glycogen, 49, 51
Glycolysis, 23, 39, 40, 54, 114
Glycoproteins, 3, 50, 54, 200
β-Glycoprotein, pregnancy-specific, 39

INDEX

Golgi zones, 110, 111
Graft rejection, 13, 148
Graft-versus-host reaction, 16, 138, 149
Granulocytes, see Polymorphonuclear leukocytes
Granuloma, BCG-induced, 145, 147
Growth, 5, 6, 32, 36, 42, 183
Growth hormone, 51, 54, 126
Gut microflora, 46, 125

Halide-myeloperoxidase system, 24, 114
Hassall corpuscles, 68, 70
Health surveys, 1, 2, 6, 7
Height, 34, 36
Hemoglobulin, 37
Hemolytic complement activity, 104, 105
Hepatic damage, 108
Hepatitis, 42, 149
Hepatitis antigen, 42, 43, 98
Herpes virus, 6, 42, 43, 125, 181
Hexose monophosphate shunt, 24, 116
Hibernation, 131
Histiocytosis, familial lipochrome, 110
Histoplasmosis, 57
Hookworm, 63, 91, 92
Hormones, infection and, 53
Hospitals, malnutrition in, 182
Host defense
  depression of, 56
  effect of nutritional deficiency on, 67
  immunological processes and, 11, 13
  in infection, 56
  mechanisms of, 11
  nonspecific, 11, 13
Humoral immunity, 13, 61, 89
Hydrogen peroxide, 24, 114
Hyperglycemia, 51
Hypersensitivity, See Allergy; Delayed hypersensitivity
Hypoglycemia, 51, 55

Immune complex, 12, 60, 61, 106, 116, 182
Immune response
  development of, 12, 13
  impairment of, 29, 56, 80, 85, 97, 101, 116
  to infection, 56
  influence of nutritional deficiency on 67, 143, 148, 159, 163, 167, 171
  protein deficiency and, 132
  vitamin deficiency and, 152
Immune system, 11, 13

Immunity, See Antibodies; Cell-mediated immunity; Host defense; Immunoglobulins; Polymorphonuclear leukocytes
Immunization
  effects of, 1, 2, 57, 59, 199
  response to, 84, 97, 101, 103, 192, 199
  schedule, malnutrition and, 193
Immunocompetence
  impairment of, 4, 29, 57, 64, 80, 96
  infection and, 56
  nutritional deficiency and, 67
  therapy to increase, 194
Immunoconglutinin, 27, 108, 109
Immunodeficiency, 4, 29, 80, 85, 97, 101, 116
Immunoglobulin(s)
  aggregated, 17
  biological functions of, 18
  classes of, 18
  development of, 16
  levels of
    energy-protein undernutrition and, 90, 97
    parasitic infestation and, 62
  metabolism, malnutrition and, 91, 95
  production by B-lymphocytes, 20
  on surface of lymphocytes, 13, 17
  structure of, 20
  subclasses of, 3, 21
Immunoglobulin A (IgA)
  antibodies, 103
  deficiency in, 104
  functions of, 18, 19
  infection and, 62, 63, 90, 91
  levels, malnutrition and, 90, 91
  physico-chemical properties of, 18, 19
  subclasses of, 21
  See also Secretory IgA
Immunoglobulin D (IgD)
  functions of, 18, 19
  levels
    infection and, 62, 91
    in malnutrition, 91, 92
  physico-chemical properties of, 18, 19
  synthesis of, 13
Immunoglobulin E (IgE)
  effect of
    on immune response, 3
    on lymphocyte proliferation, 88, 92
    on T lymphocyte rosetting, 88, 92
  functions of, 18, 19

Immunoglobulin E (IgE) (*cont'd*)
  levels
    cell-mediated immunity and, 92
    cystic fibrosis and, 61
    factors regulating, 92
    infection and, 62, 63, 91
    malnutrition and, 90, 91
    normal, 18
    parasites and, 19, 62, 91
  physico-chemical properties of, 18, 19
Immunoglobulin G (IgG)
  antibodies
    affinity of, 141
    to food antigens, 103
  fraction, transfer factor activity in, 195
  functions of, 18, 19
  levels
    in cord blood, 20, 21
    in fetal malnutrition, 93, 95
    infection and, 62, 63
    in malnutrition, 90, 91
    normal, 20, 62, 91
    parasites and, 91
  on lymphocyte surface, 17
  macrophage receptors, 25
  metabolism of, 63, 91, 95
  neutrophil receptors, 23
  physico-chemical properties of, 18, 19
  subclasses of, 20, 21, 93, 94
  surface receptors for, 13, 17, 23, 25
  synthesis of, 20, 63, 95
Immunoglobulin M (IgM)
  antibodies, 19
  functions of, 18, 19
  levels of
    in cord blood, 5, 20
    infection and, 62, 63
    in malnutrition, 90, 91
    normal, 62, 91
    parasites and, 62
  on lymphocyte surface, 17
  physico-chemical properties of, 18, 19
  subclasses of, 21
  synthesis of, 13
Immunopotentiation, 195
Infants, low birth weight, 38, 51, 184
  See also Fetal malnutrition
Infection
  acute phase reactant glycoproteins in, 3, 49
  complement defects and, 29

Infection (*cont'd*)
  diagnosis of, 201
  effects of
    on albumin levels, 48
    on bacterial killing by neutrophils, 3, 64, 119
    on carbohydrate metabolism, 51
    on cell-mediated immunity, 57
    on chemotaxis, 66, 113
    on complement system, 3, 105, 106, 108, 109
    on delayed hypersensitivity, 3, 57
    on immune response, 2, 3, 29, 56
    on lipids, 52
    on minerals, 52
    on nutritional status, 2, 4, 47, 49
    on polymorphonuclear leucocytes, 63
    on protein metabolism, 50
    on vitamins, 53
  hyperalimentation and, 182
  nutritional consequences of, 47, 49
  resistance to, 13
  susceptibility to, 1, 42, 181
  severity of, 181
Infectious mononucleosis, 57, 59
Inflammatory response, 3, 66, 112, 145
Influenza, 57, 65, 86, 98
Insulin, 51, 54
Interferon
  functions of, 13, 43
  mechanism of action of, 125
  synthesis of, malnutrition and, 125
Intergenerational effects, nutritional deficiency and, 159, 165, 170, 185
Intestinal microflora, 46, 125
Intracellular bacterial killing
  infection and, 65, 120
  in malnutrition, 117, 119, 120
  metabolic events associated with, 22, 24, 118
Iodination, 114, 115
Iodine, deficiency of, 37
Iron, 37, 49, 52
Iron-binding proteins, 123
  See also Lactoferrin; Transferrin
Iron deficiency
  antibody response in, 101
  candidiasis and, 6, 45
  cell-mediated immunity in, 77, 85, 87
  DNA synthesis in, 86, 87
  delayed hypersensitivity in, 85

INDEX

Iron deficiency (cont'd)
  diagnosis of, 37
  infection frequency and, 7, 45
  lymphocyte proliferation in, 86, 87
  lymphocyte number in, 77
  parasitic disease and, 173
  polymorphonuclear leukocyte function in, 87, 119
  susceptibility to infection in, 5, 7, 45, 174
  thymus in, 173
Isohemagglutinins, malnutrition and, 96

Kallikrein-kinin system, 51
Keyhole limpet hemocyanin
  antibody response to, 99
  delayed hypersensitivity to, 82
Kinins, 51
Kwashiorkor
  antibody response in, 97, 101
  causes of, 31
  cell-mediated immunity in, 80, 85, 87
  clinical features of, 32
  complement system in, 104
  delayed hypersensitivity in, 80, 82
  γ-globulins in, 89
  immunoglobulins in, 91
  incidence of, 1
  infections and, 1, 3
  lymphocyte number in, 75, 76
  lymphoid tissues, 67
  polymorphonuclear leukocytes in, 110
  thymus in, 68, 70, 74
  transferrin level in, 124
  See also Malnutrition; Protein deficiency

Lactate, 115
Lactoferrin, 13, 66, 123
Lazy leukocyte syndrome, 110
Leprosy, 43, 57
Leukemia, 43
Leukocyte number, 57, 63, 74, 110
  See also Lymphocytes; Polymorphonuclear leukocytes
Leukocyte endogenous mediator, 54
Levimasole, 196
Lipids, 52, 150, 152
Lipopolysaccharide, 27
Lipotropic factors, deprivation of, effect of
  on antibody formation, 163, 165, 167

Lipotropic factors (cont'd)
  on cell-mediated immunity, 163, 164, 166
  on lymphoid tissues, 165, 167
  on susceptibility to infection, 168
Liver
  damage to, 108
  enzyme activity in, 50, 51
  protein synthesis by, 26, 50
  uptake of nutrients by, 49
Local immune system, see Secretory IgA
Lymph node, 67, 71, 72
Lymphocyte(s)
  differentiation of, 12, 13
  production of interferon by, 125
  subpopulations, 15, 75, 200
  See also Cell-mediated immunity; B lymphocyte; K lymphocyte; T lymphocyte
B lymphocyte(s)
  antibody production by, 16, 17
  development of, 12, 17
  differentiation of, 13, 16
  effect of Bursa of Fabricius on, 13, 16
  function of, 12
  identification of, 14, 15, 17
  influence of T lymphocytes on, 28
  number in
    energy-protein undernutrition, 78
    health, 78
    infection, 61
    kwashiorkor, 78
  See also Immunoglobulins
K lymphocytes, 13, 80
T lymphocyte(s)
  cytotoxic, 13, 80
  damage to, 60
  development of, 12
  differentiation of, 13
  function of, 13
  identification of, 14
  interactions with other cells, 28
  number in
    energy-protein undernutrition, 75, 78
    fetal malnutrition, 76
    health, 58, 78
    iron deficiency, 77, 86, 87
    kwashiorkor, 75
    measles, 58
    old age, 186

T lymphocyte(s) (cont'd)
  proliferation of, 85
  rosetting, 14, 15, 88
  stimulation of
    PHA-induced, 59, 85, 87
    antigen-induced, 85, 86
  subpopulations of, 13, 78, 80
  suppressor, 13, 60, 79
  See also Cell-mediated immunity; Delayed hypersensitivity
Lymphocytic chorio-meningitis, 7
Lymphoid organs
  age and, 185
  in nutritional deficiency, 67
  atrophy of, 8, 67
Lymphoid tissue, gut-associated, 13, 16, 71, 73, 74, 101
Lymphokines, 13
Lymphoma, 43, 104, 189
Lymphopenia, 75, 85
β-Lysins, 126
Lysosomes, 23, 64
Lysozyme
  in infection, 66
  intraneutrophilic, 24, 66, 122
  in malnutrition, 122
  plasma levels of, 66, 122
  urinary, 122

Macrophage, 24, 63, 113, 138, 141
Magnesium, 27, 52
Malabsorption, 3
Malaria, 8, 44, 57, 92
Malignancy, 9, 29, 187
  See also Tumor
Malnutrition
  antibody response in, 97, 101
  cell-mediated immunity in, 80, 85, 87
  clinical features of, 32, 33
  complement system in, 104
  delayed hypersensitivity in, 80, 82, 84
  diagnosis of, 31, 34
  gut microflora in, 125
  hormones and, 126
  immunoglobulin levels in, 91
  incidence of, 1
  infection and, 1, 3
  interferon in, 125
  isohemagglutinins in, 96
  lymphocyte number in, 75, 76
  lymphoid tissues in, 67
  lysozyme in, 122

Malnutrition (cont'd)
  polymorphonuclear leukocytes in, 110
  thymus in, 68, 70, 74
  tissue integrity in, 126
  transferrin level in, 124
  See also Anemia; Energy-protein undernutrition; Fetal malnutrition; Iron deficiency; Kwashiorkor
Manganese, 176
Marasmus
  antibody response in, 97, 101
  causes of, 31
  cell-mediated immunity in, 74, 80, 85
  cellular metabolism in, 39
  clinical manifestations of, 33
  complement system in, 104
  delayed hypersensitivity in, 80, 85
  development of, 31
  diagnosis of, 32, 34
  immunoglobulin levels in, 91
  incidence of, 1, 31
  infection and, 1, 41
  lymphocyte number in, 78
  lymphocyte proliferation in, 85
  thymus in, 68
  See also Energy-protein undernutrition; Malnutrition
Marasmic-kwashiorkor, 32
Maternal malnutrition, effects of
  on fetal growth, 183
  on immune response of offspring, 77, 93, 95, 159, 165, 170
  on leukocyte metabolism, 39
Measles
  antibody response to, 98, 99, 103, 193
  cell-mediated immunity in, 57, 58
  effects on
    albumin level, 48
    delayed hypersensitivity, 57, 58
    growth, 6
    polymorphonuclear leukocytes, 66
  elevated cortisol levels in, 60
  infection with, 3, 42, 125
  number of T lymphocytes in, 58
  severity of, 182
Meningitis, 57
Metabolic activity
  of macrophages, 25
  of maternal leukocytes, 39
  of polymorphonuclear leukocytes, 23, 66, 113, 118
Metabolic response, infection and, 49, 51, 53

INDEX

Methionine, deficiency of, 162
3-Methylhistidine, 50
Microbes, killing of, 22, 24, 63, 87, 116, 119
Microbial products, effects on immune response, 3
Microphage, *see* Polymorphonuclear leukocyte
Minerals
　deficiencies of, 172
　response during infection, 49, 52
Mitogen, *see* Phytohemagglutinin
Mitosis, depression of, malnutrition and, 85
Monilia, *see* Candida albicans
Monocyte, 24, 63, 113, 138, 141
Morbidity, infection and, 5, 181
Mortality, malnutrition and infection-related, 1, 5, 41, 181
Mucosal infections, susceptibility to, 103
Muramidase, *see* Lysozyme
Mycobacterium tuberculosis, 44, 133
Myeloperoxidase
　deficiency of, 66, 110, 114
　iron deficiency and, 174, 175
　-hydrogen peroxide-chloride system, 24, 114
Myosin, 201

NADH oxidase, 114, 115
NADPH oxidase, 114, 115
α-Naphthyl acetate esterase, 15
Neutrophils, *see* Polymorphonuclear leukocytes
Newcastle virus, 125
Niacin, deficiency of, 37, 158
Nitroblue tetrazolium, reduction of,
　energy-protein undernutrition and, 115
　infection and, 65, 66, 115
　kwashiorkor and, 115
　iron deficiency and, 87
Nonspecific host factors, 13
Null cells,
　number of, 78
　functions of, 79, 80
Nutrients
　intake of, 40
　metabolism of, malnutrition during pregnancy and, 39
Nutrition
　assessment of, 34, 37
　immunocompetence and, 3, 5, 7, 67
　infection and, 3, 5, 47, 49

Nutritional supplementation, 7, 76, 107, 118, 120
Nutritional variables, immunological variables correlated to, 67

Obesity, 121, 177, 196
Ontogeny, 12, 16, 20, 22
Opsonization, 13, 27, 104
Otitis, 44
Overnutrition, 121, 177, 196
Oxygen, metabolism of, 24, 66, 144, 118

Pancreatic atrophy, malnutrition and, 46
Pantothenic acid, 45, 101
Paracortical areas, depletion of, 71, 72
Parasites,
　antibody response to, 98
　effects of
　　on antibody response, 63
　　on immunoglobulin E, 63, 91
　infection with, 8, 42, 44
　See also Ascaris; Hookworm
Parenteral hyperalimentation, infection associated with, 182
Paronychia, 182
Pasteurella
　antibody response to, 97
　infection with, 131
Peroxidase, 64
Peyer's patches, 16, 74, 101
Phagocytes, 13, 21, 63, 110
　See also Macrophages; Polymorphonuclear leukocytes
Phagocytic vacuoles, 23, 112
Phagocytosis
　activation of hexose monophosphate shunt during, 24
　complement system and, 25, 27, 104
　energy for, 24
　energy-protein undernutrition and, 116
　lactate producing during, 24
　by macrophage, 25
　by polymorphonuclear leukocytes, 22, 116
Phosphoenolpyruvate, 39
Phosphogluconate dehydrogenase, 114
Phosphyoglycerate kinase, 114
Phosphorus, 52, 127, 176, 183
Phytohemagglutinin (PHA)
　blastogenic effect of, 85
　delayed cutaneous hypersensitivity to, 82

Phytohemagglutinin (PHA) (cont'd)
-induced DNA synthesis, 85
-induced mitogenesis, 85
response to
  energy-protein undernutrition and, 85
  folic acid deficiency and, 86
  in iron deficiency, 86, 87
  in kwashiorkor, 85
  lipotrope deficiency and, 163, 166
  protein deficiency and, 144
Plaque-forming cells, in spleen
  caloric restriction and, 170, 171
  lipotrope deficiency and, 163, 167
  protein deficiency and, 145, 148
Plasma
  complement degradation products in, 106
  inhibitors of lymphocyte function, 60, 88
  lysozyme activity of, 122
Plasma cells, 13, 17, 20
Plasmapheresis, 133
Pneumococcal polysaccharide antigen, antibody response to, 97
Pneumococcus, 134
Pneumocystis carinii infection
  in kwashiorkor, 44
  in malignancy, 45, 190
  protein deficiency and, 44, 149, 190
Pneumonia, 6, 42, 43
Poliomyelitis virus, antibody response to, 98, 99
Polymorphonuclear leukocyte(s) (PMN)
  antimicrobial action of, 24, 26
  chemotaxis of, 23, 66, 112
  differentiation of, 22
  energy processes in, 113, 114, 118
  functions of, 27
  granulation of, 110
  infection and, 63, 119, 120
  inherited defects of, 110
  intracellular bacterial killing capacity of,
    energy-protein undernutrition and, 116, 118
    in iron deficiency, 87, 119
    in kwashiorkor, 117
    in obesity, 119, 121, 199
    in progeria, 186
    related to transferrin saturation, 87

Polymorphonuclear leukocyte(s) (cont'd)
  kinetics of, 201
  marginal pool of, 65, 112
  marrow reserves of, 65, 112
  metabolism, 23, 24, 113, 114, 118
  migration of, 23, 113
  mobilization of, 21, 65, 112
  morphology, 63, 110, 111
  nitroblue tetrazolium reduction by, 115
  number, 63, 110
  phagocytosis by, 23, 116
Polysaccharide antigens, immune response to, 86, 97
Postoperative sepsis, 182
Potassium, deficiency of, 176
Progeria, 185, 186
Properdin, 27
Prophylactic immunization, *see* Immunization
Protein
  acute-phase reactant glycoproteins, 3, 50, 54, 200
  C-reactive, 51, 126
  deficiency, effects on
    affinity of antibody, 141
    antibody production, 8, 137, 145, 147
    bactericidal function, 145
    cell-mediated immunity, 137, 144, 145, 147
    coxsackie virus infection, 150
    endocrine function, 74
    gastrointestinal tract, 101
    graft-versus-host reaction, 149
    homograft rejection, 148
    host resistance to infection, 132, 133
    immune responses, 144, 145
    interferon production, 150
    macrophages, 142, 145
    mobilization of neutrophils, 146
    parasitic infestation, 150
    phagocytosis, 138
    reticuloendothelial function, 138
    susceptibility to infection, 149
    thymus, 145, 150
    See also Kwashiorkor
  heterologous, antibody response to, 98
  iron-binding, 123
  metabolism, alterations of during infection, 50, 136
  requirements, 130

INDEX

Protein-calorie malnutrition, *see* Energy-protein undernutrition; Kwashiorkor; Marasmus; Protein deficiency
Protein-losing gastroenteropathy, 48, 60
Pseudomonas polysaccharide, effect on mobilization of PMNs, 112
Pseudotuberculosis, 131
Purified protein derivative (PPD), delayed hypersensitivity to, 57, 59
Pyridoxine deficiency
  diagnosis of, 37
  effects on
    antibody formation, 100, 133, 158
    cell-mediated immunity, 159
    nucleic acid synthesis, 159
    susceptibility to infection, 5
Pyruvate, 39
Pyruvate kinase, 39, 114

Red blood cells, antibody response to, 98
Research needs, 199
Resistance, factors of, 11, 13
Respiratory infections, 5, 7
Respiratory syncytial virus, 65
Reticuloendothelial system, 136, 141, 143, 146, 176
Rhinotracheitis, infectious bovine, 127
Riboflavin, deficiency of, 37, 53
RNA polymerase, 39
Rickettsia, 8, 42, 99
Rosette-forming lymphocytes, 13, 15
  See also B lymphocytes; T lymphocytes

Salmonella
  antibody response to, 63, 64, 96, 97, 193
  infection with
    incidence of, 44
    iron status and, 174, 175
    obesity and, 180
    protein intake and, 135, 136, 138, 139
    vitamin deficiency and, 152, 154, 155
Sandfly virus, 65
Sarcoma, 7
Scarlet fever, 57
Schistosomiasis, 57, 155
Secretory IgA antibody response
  energy-protein undernutrition and, 99, 101, 103

Secretory IgA antibody response (*cont'd*)
  iron deficiency and, 99
  reduction in, significance of, 103
Secretory IgA, levels of, 101, 102
Selenium, 176
Sepsis, 42, 66, 182
Septicemia, 43, 181
Serum antibody response, *see* Antibody(ies)
Serum complement, *see* Complement
Serum immunoglobulins, *see* Immunoglobulins
Serum transferrin, *see* Transferrin
Siderophilin, *see* Transferrin
Skin, 13
Smallpox, 12
Socioeconomic status, malnutrition and, 2
Spleen
  atrophy of, 67, 145
  germinal centers in, 71
  lymphocytic depletion of, 71, 73
  plaque-forming cells in, 145, 148, 163, 167, 170, 171
  size of, 67, 71
  stimulation of cells in, 161, 166
  weight of, 67, 71
Staphylococcus, 116, 117, 120, 182
Starvation, 7, 8, 85, 106, 119
Stem cell, 12, 13
Steroids, *see* Corticosteroids
Streptococcal antigens, immune response to, 81, 82, 86
Stunting, in classification of malnutrition, 36
Surface receptors, 14, 17, 23, 25, 60
Surgery, 29, 182
Synergism of infection and malnutrition, 2, 4, 8, 131, 132
Syphilis, 57, 59

Tetanus toxoid, antibody response to, 94, 97, 193
Therapeutic regimen, effect on
  antibody response, 99
  bacterial killing by polymorphonuclear leukocytes, 117
  complement system, 106, 107
  T lymphocytes, 76
Thiamin, deficiency of, 37, 53
Thymidine, uptake of, 79, 87, 146, 161, 163, 166

Thymolymphatic system
  atrophy of, 42, 67
  functions of, 12
Thymus
  absence of, 14
  atrophy of, 42, 68, 69
  cells dependent on, differentiation of, 14
  corticosteroid uptake in, 74
  development of, 14
  in fetal malnutrition, 70
  weight of, 67, 68, 70
Thyroxine, 126
Tissues, changes in, 125
Tobacco mosaic virus, antibody response to, 98
Tonsils, effect of malnutrition on, 71, 73
Transfer factor, 195
Transferrin
  antibacterial function of, 123, 124
  affinity of antibody to, 141
  deficiency of, 37
  levels of
    delayed hypersensitivity and, 83
    in energy-protein undernutrition, 37, 124
    in kwashiorkor, 124
    in marasmus, 124
    prognosis and, 124
  as part of host defense, 13
Treatment, 76, 99, 106, 195
Tryptophan deficiency, macrophage function in, 141
Tuberculin, delayed hypersensitivity to, 57, 59, 81, 83, 84, 192
Tuberculosis, 5, 8, 44, 57, 60, 133, 181
Tumor
  allogeneic, 8
  antibody response to, 99, 143
  autochthonous, 8
  host resistance to, dietary factors and, 7, 104, 143, 187
  syngeneic, 8
Typhoid fever, 44, 50
Typhoid vaccine, antibody response to, 96, 97, 193

Umbilical cord blood, immunoglobulin levels in, 18, 20, 21, 93, 95

Undernutrition, see Anemia; Energy-protein undernutrition, Kwashiorkor; Marasmus; Protein deficiency; Vitamin deficiency
Uric acid, 51, 54
Urine,
  biochemical analysis of, nutritional assessment and, 37
  lysozyme excretion in, 122

Vaccination, see Immunization
Vaccinia, 7
Viral infection,
  cell-mediated immunity in, 57, 59
  incidence of, malnutrition and, 7, 42, 43
  polymorphonuclear leukocyte function in, 65
  secretory antibody response to, 101, 103
Visceral movements, 13
Vitamin A
  deficiency of, 2, 43, 53, 127, 131, 132, 153, 157
  excess of, 180
Vitamin B_1, deficiency of, 37, 53
Vitamin B_2, deficiency of, 37, 53
Vitamin B_6, See Pyridoxine
Vitamin B_{12}, deficiency of, 37, 53, 86, 161, 163, 188
Vitamin C, deficiency of, 132, 170
Vitamin D, deficiency of, 45, 132, 156
Vitamin E, deficiency of, 156

Wasting, in classification of malnutrition, 36
Weanling diarrhoea, 46, 125
Weight, 6, 34
Weight-for-height, 34, 36
Western equine encephalomyelitis, antibody response to, 98
Wet-tail, 131
White blood cells, See Leukocytes, Lymphocytes, Monocytes, Polymorphonuclear leukocytes

Xerophthalmia, 53

Yellow fever vaccine, response to, 98, 193
Yolk sac, 12

Zinc, 52

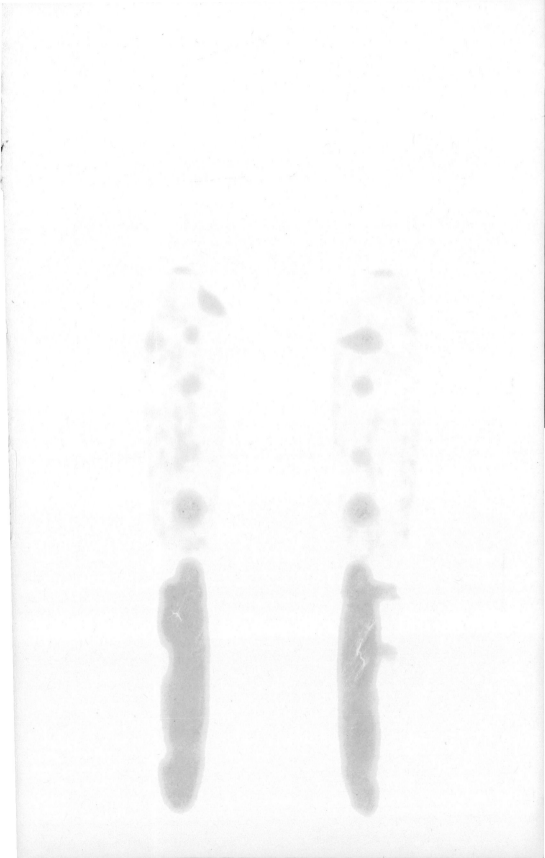

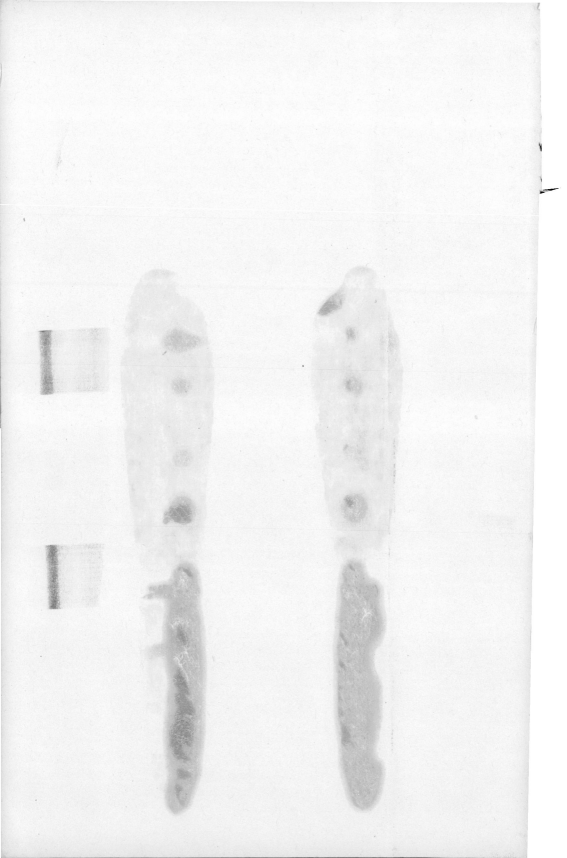